RICHARD ROBINSON is the author of twenty books on popular science, including the Science Magic series (Oxford University Press), which was shortlisted for the Aventis Prize

Robinson), which he

He works full-time as a scie

Science Festival and regularly performs demons

the world from Boston to Beijing. In another life he also co-founded the busker's pitch in Covent Garden and *Spitting Image*.

My Manager
and
Other Animals

Evolution and Survival in the Office Jungle

Richard Robinson

Constable • London

Constable & Robinson Ltd
55–56 Russell Square
London WC1B 4HP
www.constablerobinson.com

First published in the UK by Constable,
an imprint of Constable & Robinson, 2014

A copy of the British Library Cataloguing in Publication Data
is available from the British Library

ISBN: 978-1-47210-667-4 (paperback)
ISBN: 978-1-47210-672-8 (ebook)

1 3 5 7 9 10 8 6 4 2

Printed and bound by CPI Group (UK) Ltd, Croydon, CR0 4YY

To Morgan and Georgia

ACKNOWLEDGEMENTS

It took five years to write, and a lot of sage and knowledgeable people were enlisted along the way; my apologies for missing out many of them, including the geniuses of Wikipedia and the internet. When I was first setting out, and it was looking like a loose collection of underweight quips with overweight analysis, it was Nabil Hamdi who said, 'Have you come across *Emergence*?' and started giving form to the book and to my life. All else followed naturally. Thank you Jonathan Bacon for helping organise a day devoted to emergence and Nic Stacey and Jim Al-Khalili, whose TV series *The Secret Life of Chaos* explained how chaotic organisations manage to survive and even thrive; to Lucy Kellaway of the *Financial Times*, Berry Winter and James Hammerton-Fraser, Paul Levy and my brother Philip among others who gave me stories and insights. All at the Brighton and Hove Chamber of Commerce and the Catalyst Club who have let me practise the material on them over the years. And of course Leonie Bennett, for patiently reading a dozen versions of the book over half a decade.

CONTENTS

INTRODUCTION

My manager

Aren't we humans clever!

If we measure cleverness of a species by how much it can spread into every nook and cranny on Earth, however cold, hot, dry or sticky; can exploit every animal, vegetable and mineral it finds there; can create light when it's dark and heat when it's cold; can invent a device the size of a chocolate bar and use it to talk to someone on the other side of the world; can explore other galaxies while contemplating the meaning of existence, then the human species is by far the most imaginative, enterprising and brilliant the world has ever seen.

So why is my manager such a dork? How much time does he spend advancing the glorious species, and how much in pointless list-making, useless meetings and petty acts of sabotage on everyone around him? Doing the wrong deals with the wrong companies in the wrong countries at the wrong time, taking wrong advice from wrong people. Why is everything over-hyped, overdue and over budget? It's chaos here. It just won't do. I'm a brilliant jewel of evolution, get me out of here!

Other animals

Two and a half thousand years ago Aesop, a Greek slave[1], told tales about animals, designed to tell us a little about ourselves. He is credited with a number of fables now collectively known as Aesop's Fables. A satirist, he used animals as a mask for his jibes against the cheats, freeloaders, bullies and snake-oil salesmen who were his real targets. Now we have scientific evidence that Aesop was right: animals aren't just simple bags of blind reflexes, as we thought. It turns out the animal kingdom is as full of cheats, freeloaders, bullies and snake-oil salesmen as our own. We evolved from them, after all, so why not? But also their subtle minds feel sorrow, remorse, love, compassion, scorn, can read each other's intentions, make plans, invent tools ... In fact, just about everything we can do they can do except, possibly, cryptic crosswords.

We don't like to hear this. Our brains are our pride and joy. They are what sets us apart from lesser animals. (Lesser animals, aside from their brains, are almost all stronger, cuter, neater, more practical or better dressed than we are). Willingly we go through the pain of giving birth to such a huge brain and the unforgiving skull around it because we treasure it. It can do such amazing things. It invented central heating and aeroplanes, for instance, which no other animal has.

In 2009, researchers at Cambridge tested one of the best known of Aesop's fables directly.

> A crow, half dead with thirst, came upon a pitcher which had once been full of water; but when the crow put its beak into the mouth of the pitcher he found that only very little water was left in it, and that he could not reach far enough down to get at it. He tried, and he tried, but at last had to give up in despair. Then a thought came to him, and he took a pebble and dropped it into the pitcher. Then he took another pebble and dropped it into the pitcher. Then he took another, and another, and so on. At last, at last, he saw the water mount up near him, and after casting in a few more pebbles he was able to quench his thirst and save his life. (Aesop)

In a laboratory, the appropriately named Christopher Bird gave some Caledonian crows the same problem – in their cage was a glass tube (the 'pitcher') with a little water in, and a grub floating on the top. The grub was too far down in the tube for the crow to reach with its beak, but a small pile of stones was nearby. After a few tries the crow worked out what to do, and dropped the stones into the tube until the water was the right height to pick out the grub. Even more remarkable, it learned to pick the larger stones first to get the quickest result. So an animal with a brain no bigger than a walnut managed to solve a problem which would have defeated some of those around me at work.[2]

Our tour of the animal kingdom will give choice examples of 'human' traits. But most importantly it will come back to

RICHARD ROBINSON

a special pair of animals again and again – the ant and the ape – because they symbolize two particular character traits that create the chaos of our lives: firstly the deep desire to flock together with everyone else, like ants, secondly the compulsion to argue with everyone once we get there, like apes. We need both of these impulses to work together in antagonistic harmony. ('Antagonistic harmony' is a phrase used to describe how muscles – the biceps and triceps, for instance – work together to make a limb such as the arm function, with one pulling to bend it, while the other is needed to flex it).

Antagonistic harmony

Antagonism is the watchword in the workplace nowadays. Business analysts see work through the eyes of apes. The office jungle throbs with rivalry. Young apes challenge the alpha ape for dominance and compete against each other for promotion. It's nature red in tooth and claw as we bite and scratch our way up the pecking order. And we don't do it for anyone but ourselves. It's Me Me Me. Me beating the other applicants to a job. Me fighting my corner at work. Me hitting my targets. Me getting promoted to another department, where Me finds new rivals to fight. In fact, evolutionary biologists say, our lives are ruled by 'selfish genes', which calculate the benefits to Me before Me does anything for anyone else.

Harmony is the opposite; an instinct for joining and cooperating. It makes us help our colleagues rather than put one over on them, listen to them rather than boss them, empathise with them rather than brush them aside. This is the spirit of ants, who are such compulsive harmonizers that they can't function on their own, and need a cluster of at least a hundred to get the slightest thing done.*

These two, antagonism and harmony, were big on the day years ago when I was working for a design company. The boss was off ill, bequeathing his team a deadline for a brochure, but left no more instructions than that. We needed to decide together on text, pictures, fonts and layout. And soon. The boss's choice team handily comprised a designer, writer and graphics specialist, though in the course of a hectic day we found that the writer had no idea about spelling, grammar or syntax. Fortunately we discovered the artist was a bit of a whizz at writing, which was just as well, since she couldn't draw, a skill which the designer excelled at. The team thus inverted its skill set and just about squeezed the brochure out on time, each one of us pleased to be using a skill other than the one we were being specifically paid to do, and quietly wondering why we had never been asked before. The actual brochure-related work took minutes; the arguing filled the rest of the day. It was friendly arguing, because we knew where we needed to get to in the end; it was intense, because we all had opinions on how to get there. The end was harmonious, the path was antagonistic, but as a result of the struggle we bonded in a way we couldn't have if we had gone the easy path.

---

* Richard Dawkins, geneticist and author of *The Selfish Gene*, points out that co-operative behaviour may well originate from selfish intentions, deep down in the genome. My genes want to be protected, and if behaving altruistically helps, this is what I'll do. Whatever the merits of that argument, the outward appearance is cooperative behaviour.

Section 1 shows that the ant is big in us. The forces of harmony bring us all together in ant–like swarms. We are built to cooperate.

Section 2 sets out in contrary style to establish the opposite; that we are antagonistic, aggressive, hierarchical and just generally cussed, like apes. Having proved that both are true, we will then see how they collide, as collide they must. We find that pure antagonism doesn't work on its own, but neither does pure harmony. The only way forward is with both working together in antagonistic harmony. We'll be looking at the chaos this creates and seeing why chaos is actually a good idea.

Section 3 helps you negotiate the cultural rapids at work. Antagonistic harmony may be the answer, but it produces some pretty crazy questions along the way. Understanding the causes of dysfunctional organizations can help one to negotiate the chaos calmly.

Section 4 raises the paradox, that antagonistic harmony, though it is chaotic, is the path to the future, because it is the source of creativity.

Section 5 concludes that since we live on a chaotic planet in a chaotic universe, being able to cope with it is pretty important. In fact, as you may have already observed, we are experts in chaos.

My Manager
and
Other Animals

SECTION 1
HARMONY

CHAPTER 1

THE EVOLUTION OF HARMONY

The history of harmony

Bacteria

To discover why togetherness is a good idea, we must travel back in time, way back four billion years to when the freshly formed Earth was in the middle of what is known as the Hadean Age; in other words, like Hades, with fire and brimstone, meteorite bombardments and cataclysms being delivered with the regularity of the postman. From that hell emerged the organized chaos which was life, in the very simplest form – bacteria. Right from the off, bacteria discovered that sticking together meant they were going to be winners. By 'sticking together' I mean just that: they secreted a polysaccharide 'glue' which bound them to each other and to the rocks they sat upon. By being bonded together they were protected from predation and there, stuck to the rocks, the nutrients in the ocean bathed them in everlasting lunch.

The next generation set up home on top of their parents. Each generation built on the heads of the last. The rock grew slowly bigger, layer by layer. They are still doing it today in some parts of the world. They are called stromatolites. In Shark Bay, Australia, the shallow, nutrient-rich sea is crowded with both them and tourists who gaze in awe at archaean life, still vibrant after four thousand million years.

If you want to see for yourself how biofilm works you don't have to go back four billion years, or even to Australia. Take a look in the mirror. Smile. The 'plaque' which coats your teeth is biofilm. Oral bacteria have evolved to stick to your teeth so they can never be far away from the sugar they love so dearly. They are so successful that a whole industry has been created to try to hold them back. Modern dentistry shows us that the old saying, 'United we stand, divided we fall' has a four-billion-year provenance.

Starlings
Harmonizing is hard-wired. Out in the wild it's 'harmonize or die'. Later on we'll see how starlings 'worked out' that to try to reach the centre of the flock is a good idea, since those who didn't care, who flew blithely to one side, flitting prettily and showing off their barrel roll, were skilfully snatched by the peregrine falcon flying alongside looking for lunch. Only those who felt an irresistible urge to cluster in the centre survived to the next generation.

Humans
The same is true for humans. In the past, when times were tougher than today, people who decided to live outside the

tribe, or who were expelled, were in danger from not just wolves and bandits, but disease and parasites, with nobody to help deal with them. As they were picked off, the ones that remained were the fraternizers, the ones who evolved a yearning to be part of the tribe.

Our ancestors clustered together in towns that could defend themselves against invaders (as well as clustering together at other times to launch invasions of their own). The towns grew into cities. In which groups of businesses hung out together. As long as 5,000 years ago, clusters of shops could be found in the centre of Uruk, the world's first city, on the banks of the Euphrates. In London before the sixteenth century, the clustering had gone one stage further; a single trade could take over entire streets. Although the Great Fire of 1666 swept the shops away, the street names are still there: Ironmonger Row, Haymarket, Fish Street Hill, Baker Street and Pudding Lane (where the Great Fire of London began, perhaps in an oven full of puddings). Nowadays the clustering goes on, though in different streets. You won't find bonds in Bond Street, but you will find art galleries, lots of them. Go to Hatton Garden today for jewellery (but not hats). No needles or thread in Threadneedle Street, but if you want, you can get stitched up by one of the many finance houses that have clustered there.

In spite of the higgledy-piggledy growth of these places, great things can happen very fast. Between 1700 and 1850, Manchester grew from a large village to be the powerhouse of the British Empire, with a population of 300,000, and all without more than the most vague attempts at central coordination. It wasn't even recognized as a city by central government until

1853. Everything was done through local deals and arguments, weaving among those of the wider neighbourhood.

Nowadays, corporation headquarters cluster in city business centres. We can see instinctive clustering in their marketing, too. They sprout names in harmony. In the 1920s there was a fad for giving products names ending in '-o', thus Aspro, Glaxo, Oxo, Omo, Zippo, Vimto. Brasso, Bisto; everybody did it. More recently the clustering was around Latinate names.

Look at some of our big multinationals; Invensys, Unisys, Infosys, Novartis or Inventis. When a company spends millions of pounds on researching a distinctive new name, then opts for one which sounds like all the others, you have to believe in the strength of the harmonizing instinct.

Through history we find humanity is a reassuringly cooperative species when there is a crisis. We have survived disaster after disaster. When tragedy strikes we all stop what we're doing and cluster together to cope with it. After an earthquake neighbourhoods gather round to dig through collapsed buildings and rescue the injured. Everyone finds a useful job to do in coordination with the rest. They form chains to clear the rubble; they run off to find stretchers, bandages or water; they all stop everything together to listen for signs of life in the rubble. It's chaotic when you're in the middle of it, yet sure enough the combined effect is survival and continuity. Here are some examples of disasters from the past:

One third of Europe died during the Black Death in the fourteenth century. Europe recovered. The only remaining sign of it is a paragraph in the history books.

The Great Fire of London destroyed one of the biggest cities in the world in 1666. You wouldn't notice today. Everyone scurried around to rebuild it.

After being flattened by allied bombing in 1944, Dresden's residents built it all over again. If you visited Dresden tomorrow you wouldn't think anything had happened.

A quarter of a million people died in a single day in 2004 when the Asian Tsunami struck. Towns were destroyed and millions of square miles of land disappeared. The whole world chipped in to help put everything back. Within five years it was business as usual. I don't want to belittle the enormity of the individual tragedies that befall us, but to praise humanity's ability to recover in the long run, thanks to our instinct for cooperative teamwork.

Wikipedia

Here is the great revelation of our day: the internet's ability to create unmanaged magic. It must have seemed like the ultimate folly when Jimmy Wales and Larry Sanger thought of Wikipedia in 2001; an encyclopaedia written by anyone and edited by anyone else. No managers, no commissioning editors; an entirely bottom-up knowledge base (Wikipedia employs one person). By contrast, Encyclopaedia Britannica is an ancient and august institution whose structure is the purest of human pyramids. When *Nature* magazine compared the two in 2005, they found an average of four errors per article in Wikipedia. Startlingly, though, they found an average of three errors per article in Encyclopaedia Britannica. So the two match up very well. The internet has opened our eyes to the extraordinary self-organizing abilities of humans. Wales and Sanger grasped the opportunity firmly with both hands – or rather with no hands. It just runs itself.*

---

* When a large group of people work to compile information on a given topic, disputes may arise. To resolve a dispute on Wikipedia, the editors will share their points of view on the article's talk page. They will attempt to reach consensus so that all valid perspectives can be fairly represented. This allows the site to be a place not only of information but of collaboration. Many users of Wikipedia consult the page history of an article in order to assess the number, and the perspective, of people who contributed to the article.

CHAPTER 2

THE SCIENCE OF HARMONIZING

To harmonize effectively you need to send and receive signals and act on them. Ants communicate, urgently and continuously. Every step they take, they drop chemical signal molecules – pheromones – on the ground behind them. Ants have at least twenty chemical pheromones they can blend in different ways to give composite signals, like letters of a chemical alphabet joining to form pheromone words. The 'words' are spread over the ground where they walk, like a combined newspaper and to-do list, though they are not read but smelt, like a dog smells a lamppost, to give them a notion of what everyone else is doing, what needs doing next, how much and where.

Human communication systems are different. Ants do a lot with smell and touch. We're rubbish at both of those. We have a lousy sense of smell and there are laws against uninvited physical contact. But we have good eyesight and pretty good hearing, so these have evolved to become our main ways of communicating. With them we communicate without knowing – we empathise.

We always knew we could empathise. We easily share our mates' sorrows and joys, or even feel a sympathetic twinge of pain when they bump their head. What we didn't know before is just how much we empathise, how much we can't stop ourselves and what a huge effect it has. The latest research has revealed the hidden signals we send and receive and how we adjust our behaviour unconsciously in response.

Mirror neurons

Kim was a manager in a busy stables. She had known her staff and their families since she was a child; she even went to the same school as many of them. When they were in difficulties, whether personal or professional, they came to Kim, knowing she would empathize with them. She admitted that she became exhausted when one of her staff was in emotional upheaval, as if she were going through it at the same time.

There could be a good scientific reason why Kim couldn't stop empathizing – mirror neurons.

In 1998, Giacomo Rizzolatti and others at the University of Parma, Italy, were investigating the age-old problem among philosophers and psychologists: which part of the brain is used to understand the thoughts of others? Rizzolatti was studying macaques (it might just as well have been MacKirks or Maclarens, since we have discovered humans show the same responses). When a macaque was performing an action – picking something up, for example – the experimenter could display on a screen the pattern of neuronal firing in the motor cortex. The instruments were so precisely tuned that slight differences in the muscles the macaque used between picking up a peanut and picking up a pencil caused differences in the firing pattern. The shock surprise of the experiment came when at one point the experimenter picked up a peanut to rearrange it. The watching macaque's neurons fired off in exactly the same way, as if he, (the macaque), was actually doing it. The macaque's mind mirrored the experimenter's. Further experiments proved that large areas of the brain fire off in this way.[1]

With what we know already about empathy and mimicry in our daily lives, we should hardly be surprised about this, but here might be the proof. Mirroring is compulsory. We automatically create in our minds the neural pattern of the person we are communicating with, whether or not we decide to consciously copy them. This reverses the traditional view of mimicry. We used to think that copying another person was something we could choose to do. We can now reckon that copying comes so naturally that *not* to imitate each other is an effort. This is our natural, inbuilt, on-board empathy machine.

Body language

Everybody has heard of 'body language' now, and there are teach-yourself books, dictionaries and checklists available everywhere that map out the signals we transmit by gesture or the way we sit. But with or without guide books, we seem to read each other's body language with great skill. Now, thanks to video technology, scientists have been able to study body language in detail. They show mirroring to be pretty rampant.

When people chat, gestures can be reflected around at the rate of two or three a second. No guide book can help you here; it's a language you learn unconsciously. The teach-yourself books will tell about the big stuff; whether a hand is open showing the palm or closed into a fist, eyes are wide open or half closed, gazes are direct or out of the corner of the eye, and so on. But those gestures are like the notes on a piano – all very well, but what counts is the tune you play on them. A good conversation is accompanied by an unconscious body-language boogie.

We can reach a much deeper level yet. Careful analysis of people having a conversation reveals tiny sprites – movements which are too subtle to spot without watching the video frame by frame. They are so small that many psychologists deny we could possibly detect them, but when we speak of 'something funny' about somebody it may be that we are dimly aware of a micro-gesture communication below our normal threshold of perception. Is this what we mean by a 'sixth sense'?

As part of a research project, I studied video footage of a CEO visiting one of his company departments. He talked with everyone there and was utterly warm, charming and full of praise. As he turned from talking to the manager I thought I caught something on the recording. For one moment I glimpsed a different man. I had to slow the video right down, but the gesture was as clear as it was completely out of keeping: for two frames his eyes cast upwards and his face said 'What an idiot!', before he carried on turning in his utterly charming and disarming way to the next person. The micro-gesture lasted for no more than a tenth of a second, too quick

to be consciously noticed by anyone. But unconsciously? Who knows.

What we do know is that if you don't have verbal language you depend on body language and grunts. So that covers all the other animals. They read each other and they do it rather well. Starlings in their murmuration avoid bumping into each other while flying at death-defying speeds; gorillas read signals delivered with no more than an eye-blink; zebras share water holes with lions if their body language says they aren't hungry. Anything that helps the survival of the species will be evolved, including deep psychological insight.

Non-verbal verbal language

Another shock: what we say counts for very little. The research is clear. Our fond belief that language is humanity's crowning glory crumbles to dust in the spotlight of scientific analysis.

Alex Pentland and his colleagues from the Human Dynamics Group at the Massachusetts Institute of Technology studied operators in a call centre in Inverness, Scotland, using devices that monitored simply the pattern and tone of the conversation, ignoring the words themselves. They found that they could predict the ultimate success or failure of a call by listening to a few seconds of a conversation's tone. The most effective operators spoke little and listened a lot. When they did speak, their voices fluctuated strongly in amplitude and pitch, suggesting interest and concern at the needs of the client. Operators who spoke with little variation came across as too determined and authoritarian.

Researchers conclude that what you are actually saying counts for less than half of your effect. This isn't too surprising; every other animal does without language, and does very well. Apes live in complex communities with strong communication and subtle interplays without using words. Complex language in humans is a very recent invention, so it's a bit of an add-on in evolutionary terms: we should not set too much store by words and logic alone.

A supervisor of a vast tooling shed near Birmingham, where they were using mechanical drills, saws and lathes to make not just a wide variety of tools but also a vast cacophony of noise, said he could understand what everyone was thinking just by looking at them from the other side of the shed. Their stance, the speed or fluidity of their movements, all told him volumes about his workers. He could even tell whether they had had a tiff with their partners the night before or if they had financial worries.

Oliver Sacks, in his landmark book, *The Man Who Mistook his Wife for a Hat*, describes a group of patients at his hospital, who suffered from aphasia – the inability to understand the meaning of words – laughing uproariously as they watched a speech by Ronald Reagan, the Hollywood actor turned president of the USA. As Sacks points out, aphasics may not be able to understand words as such, but they can work out the meaning from the many other clues they tune in to – tone of voice, emphasis, inflection, as well as body language clues. Because their condition makes them word-blind, they have honed those other senses. The aphasics could see something the rest of us missed – that Reagan himself didn't believe what he was saying in his speech, even though he was trying to sound convincing. Reagan's skilful acting deceived the aphasics not a jot, though the speech fooled millions of other less unfortunate people.

So it turns out that when you listen to someone speaking, a lot of time is spent *not* listening to the words themselves, but analysing the tone of voice for hints of emotion, then analysing the emotions for nuance, pace and pitch. A simple phrase, like 'well done', can be inflected to sound like 'You are a genius, I could never have worked that out' or, with a slight sarcastic lilt, like 'You shambling idiot, it is clearly by the merest accident that you have stumbled on the right answer, which I worked out hours ago'. Even the areas of no sound – the pauses – are significant. When you ask somebody, 'So you've read the draft of my new novel. What did you think of it?' then there's a twenty-second silence before they say, 'It's probably the finest piece of literature I have ever set eyes on', you should consider some rewriting.

Pheromones

Although a sense of smell is never considered our strong point, there's more to it than you think. When Alan De Mare turned up to see how one of his suppliers was getting on with designing and building a security locking system for his garden shed factory, he felt straight away that something was up. As he put it, he 'smelt a rat'. The device he was shown worked fine, but there should have been 500 of them not just one. Don't worry, they said, the rest were in production.

He did worry, and they weren't. He had indeed smelt something – panic.

The smell of panic was carried by signal molecules called pheromones. Every animal on the planet transmits these. Ant pheromones we have already encountered (page 9). Dog pheromones are laid down and sniffed up whenever Rover passes a lamppost. Among humans they are emitted from hairy areas like the armpits. (If you ever wondered why evolution retained hair in our armpits, now you know: the hairs provide a large area for the pheromones to launch from, by evaporation). As signals go, pheromones are pretty revealing to all and sundry. One sniff and people could know whether you're anxious, happy, angry, menstruating, ovulating or panicking. It's likely we produce many more pheromones: scientists have found that ants use twenty and bees have at least thirty, so we might expect humans to have as many if not more. But we don't know at the moment because there isn't much research being done. This is surprising, in view of the fact that smell is such an important sense. Why no research? Either it is too complicated a subject and the science hasn't developed far enough yet, or there is something so distasteful about us broadcasting our deep dark secrets all around.

People who share offices may shave their armpits because that cuts down 'unpleasant odours'. Then they'll add odours of their own from the wide variety of differently unpleasant ones on sale in chemists, and assume that the pong problem is solved. Not so; pheromones have been our chief signalling system since we were bacteria, and they will never quite go away. Panic, trust, love, rivalry, fear, all these will make their presence felt – or smelt – in the office. But we can cope.

We have been practising for four billion years, after all. For the time being all we can say is that the pheromone mix pervades the workplace in a way too subtle for us to measure, too profound for us to understand and too complex for us to deal with.

Dissolving the self

Scientists have found a number of tricks we use to harmonize with everyone around. To put it briefly, a part of you belongs to them and a corner of them is yours.

We might like to think of ourselves as sealed units, proudly detached from those around us and packed with individuality, but our identity is not so clearly drawn as we believe. There is fuzziness at our boundary with the rest of the world. Firstly I am actually a Multi-Me. There is a fuzziness at the edge of what I call 'me'. I may feel sure of who I am when I look in the mirror, but when someone comes into the room I start to behave differently; I have a special 'me' to put on display for them. Then there's another 'me' I save for the family, and another one when I'm in a crowd, or there again at work . . . I have several 'me's to pick from.

Secondly, I am Mega-Me. My body can extend itself. When I play a computer game I become the character on the screen, and can steer and veer like them. When I park a car I become part of the car; my own 'off-side rear bumper' is as one with the car's off-side rear bumper. When it knocks the car behind I wince, not just from thinking about the repair bill, but because it really seems to hurt.

Thirdly, I am Micro-Me, since I can make space in my

mind for you; I can empathize with you. When you are happy your happiness spreads into me and I'm happy too, thanks to my mirror neurons. When you bump your head I feel your pain. So I know what you are thinking. I put myself in your shoes. This helps me to know you will understand me and work along with me on an idea. You are of course doing the same with me, so the boundary that separates us is permeable. We are merging psychologically.

So we have multiple personalities. It can be quite a task to compartmentalize the different characters. The finest example of a naturally occurring dual personality is the scientist who goes to church. Quite a few do, convinced in the truth of evolution with their mind, but in their hearts believing there must be 'something' out there.

It is hard to be consistent when there is so much fuzziness in my various 'me's. Everyone has his own style of 'self'. Think of some of the others at work; you can probably describe their character in a sentence. 'John is an anxious kind of guy', 'Julie likes to be in control', 'Jane is always jolly'.

Although you expect John, Julie and Jane to be comfortably predictable from day to day, you do also expect them to be able to flex those personalities according to circumstances. When something good happens, then you hope John will allow himself a smile. If something bad happens, you hope jolly Jane will take things a bit more seriously. If John can't lighten up ever, or if Jane keeps cracking her jokes during a crisis, they'll get some funny looks from the others. On the other hand those who flex their personalities too much can be just as disturbing. Have you met people who are so willing to empathize with you that they start talking with your accent as soon as they meet you, or finish your sentences for you? They invade your mind. Most of us are in between, with a 'me' solid enough to be individual, but soft enough to empathize with those around us.

So all this picking and swapping and flexing causes us to wonder at times who is looking back at us out of the mirror, but the good side is it allows us to do that most important thing – harmonize with everybody else.

In conclusion – mimicry

Add all these together – mirror neurons, body language, pheromones and empathy – and the result is that we copy and blend with each other, quite slavishly, with results which are at times wonderful, at times preposterous, and nearly always a little comical. A case in point is the mimicry shown by some secondary school students when they decide to challenge authority. In schools which have uniforms, including school ties, the students show their independence by defying the dress code. They do their ties up wrong, with huge knots at the top and tiny tails, instead of the small knots and long tails set down in the school's dress regulations. Comically, the same defiance is shown in every school across the nation. Their ties are worn with identical knots from coast to coast, a dress code of defiant 'individuality' which is more rigorously policed by the students than any headmaster could wish for.

Later we will celebrate the richness and variety of comedy which slavish mimicry produces. Right now we must examine the psychological climate of the shared workplace.

CHAPTER 3

EGOLOGY

We have created a new world, and we need a new science to study it. Instead of ecology, the study of natural habitats, I propose Egology, the study of the unnatural habitats of the workplace, for here the human species has built its own brand-new environment. The climate is maintained at a steady temperate level. The terrain flat, with a hardy ground cover of carpet and occasional tripping hazards. Desks rise above the undergrowth in regular outcrops and among them roam the sole inhabitants of the egosystem. No other creature but humans is permitted, unless it has been cooked, sliced and inserted in a bap. The real sustenance for the indigenous creatures, which they harvest continuously, is words. They read, write, speak, email, twitter and text among the tables and cables.

RICHARD ROBINSON

Evolution, Wevolution and Mevolution

Three forces mould this environment: evolution, wevolution and mevolution.* What is the difference between them?

Evolution we know how Darwinian evolution works – slowly. It proceeds by tiny changes to genes over thousands of years. Evolution gave us the hard-wired stuff – a body this shape, with this many legs and arms and this size and shape of brain, but it took a billion years to do it. It cannot explain the explosion of culture that we have witnessed over the past few hundred years. It cannot explain the modern office.

Wevolution can explain that. As well as our hard-wired genes, we hand on to the next generation a whole packet of genes for adaptable behaviour, which we use to cope with whatever the fickle world flings at us. For instance, you know that when it's cold we shiver, shelter from the wind and huddle together – that's genes in action. But we also use wevolution to weave clothes, create fire and invent central-heating systems. We have wevolved our own, supernormal world, rising above the natural world of earth, air, fire and water which bears down on other life forms on Earth.

Wevolution created the science that lies behind the technology that lies behind the business that is the reason you are in the job you are in. It produced our culture, religion, language, dress and fashions.* Wevolution is autocatalytic; each new development inspires more new developments, each inspiring more in exponential acceleration. Surfing the internet is for me just like surfing a real, gigantic wave; the breaking tip of an ocean of wonders, rushing up through the centuries and across continents. While evolution is slow, Wevolution is exhausting.

Mevolution is to do with you and the way you adapt to fit in with your job: how you dress, the new words you pick up, who gets copied in on emails, how many decorations you are allowed to place around your desk, when and how coffee is made, how visitors are treated, how the manager is treated.

Evolution, wevolution and mevolution interlink at many points. For example:

Evolution over millions of years gave us two legs, a sense of balance and an enthusiasm for competition.
Wevolution over a hundred years created the modern game of football.
Mevolution is the way you've learned how to dive, roll over and scream in pain, hoping to be awarded a free kick.

Evolution gave us the desire to acquire.
Wevolution gave us the salary scale and annual negotiations.

---

* Richard Dawkins calls them 'memes' to distinguish them from genes, which are instructions on the DNA itself.

Mevolution ensures that managers keep salary scales and negotiations secret, because they know there'll be an uproar if the payroll is leaked.

Evolution gave us the instinct to flock together.
Wevolution provided the water-cooler.
Mevolution has us chattering around it, and managing our reputation carefully while we're there.

Surviving in the egosystem

Dwelling in our Egosystem, we hope to avoid the ways of animals in their cold, damp ecosystems, with their uneven floors and mud. However, we can't quite help being reminded of our evolutionary roots. The primitive laws of nature keep emerging into our offices and suburbs, disguised as laws of life. The following sections outline some of them: being nice, being fair, giving gifts, dressing appropriately, joining in with fun and games, doing meetings, greetings and tweetings, and generally being seen to support the company mission, sing the company song, salute the company flag, whatever the company needs.

Being nice: the way of the elephant
Being nice is the glue. In Dale Carnegie's all-time classic 1936 book about successful deal-making, *How to Win Friends and Influence People*, the main – almost sole – advice was 'be nice'. It is as relevant today as it was seventy years ago. Below are some items on his checklist:

- Don't criticize, condemn or complain
- Give honest and sincere appreciation
- Arouse in the other person an eager want
- Become genuinely interested in other people
- Smile
- Give the other person a fine reputation to live up to
- Make the other person feel important – and do it sincerely

And so on for 260 pages.

Does the niceness rule happen with other animals? Yes, wherever animals need to live and work together they tend to be nice to each other. They're always nice to their immediate family; usually nice to relatives; quite often nice to the neighbours; a lot of the time surprisingly nice to total strangers. They're even nice to other species.

In his book, *The Beauty of the Beasts*, Ralph Helfer witnessed a baby rhinoceros getting stuck in a mud pool. Its mother was powerless to help it, and eventually had to abandon it, returning occasionally to see if anything had changed. While she was away something did change. A group of elephants came by. Seeing the baby rhino in distress, one of them knelt gingerly on the hard ground at the edge of the pool and managed to get its tusks under the rhino. As it began to lever it out of the mud, mother rhino came back. Assuming the elephant was up to no good, she charged him. He had to escape to avoid being gored. Every time mother rhino went back into the bush he tried again to lift the baby. And every time he tried again the mother rhino returned to attack him. There was absolutely no reason why the elephant should help a member of another species, particularly since he was threatened with impalement for as long as he did. Yet he did.

After the Asian Tsunami of 2005 an orphaned baby hippo was adopted by a 100-year-old giant tortoise.[1] They were as useful to each other as a tortoise and a hippo can be – which is not a lot – but they bonded because there was something going between them, something deep, and deeply strange. We know it as altruism – helping another, even though it will not be of any obvious help to you.

Jambo, the gentle giant gorilla of Jersey Zoo, shot to international news stardom overnight on 31 August 1986, when five-year-old Levan Merritt fell into the gorilla enclosure. Jambo stood guard over him, keeping the other gorillas at bay and occasionally stroking him, until the ambulance arrived.

When Koko the gorilla adopted a kitten in 1985, when Jorong the orangutan rescued a drowning chick in 2011, when dolphins have rescued swimmers in distress, whenever humans rescue animals, in all these unlikely instances we can guess there is something going on which defies commonsense. Altruism often seems to appear from nowhere for no reason, but we know the reason well enough: mirror neurons, and all the forces of empathy laid out in the last chapter. Quite simply, if we are nice to others, we are nice to ourselves by proxy. Likewise if we diss others, we diss ourselves.

Some scientists worry about this. The strictest interpretation of the 'selfish gene' theory is that we only do things that help our own genes. Any altruistic act is done only to benefit us personally, or at least our family. Rigorous adherence to the rule can make life complicated: A car crashes into a lamppost just across the street. Quick as a flash you run over, open the car door and ask the injured driver about their parents and grandparents, because if you're not related, unfortunately you won't be able to help.

Well, no, you don't unless you're a certain kind of scientist. The geneticist J.B.S. Haldane was asked if he would lay down his life to save his brother. He did a lightning calculation and concluded that no, he wouldn't. However, he would lay down his life for two brothers or eight cousins. Apparently that was a fair exchange, based on genetic closeness. Most people aren't as good at maths as Haldane. In the heat of the moment they tend to give the benefit of the doubt and act altruistically anyway. It's a very easy thing to do. So easy that we do it to anyone, even total strangers. Is this bad maths in action? Should we stop to work out the profit to ourselves?

As it happens, evolution is not that good at maths either. It operates chaotically, in rough approximations. Broadly speaking, the genial gene tends to survive better than the selfish gene. If you could separate the selfish people from the generous, then ship all the selfish to one side of the world and all the generous to the other, you'd quickly find one side making a huge, wonderful success while the other half sulked themselves to extinction. Our emotions have evolved in a way that supports this; we feel good when we are generous and bad when we are stingy. Emotions take a very long few million years to evolve, so we should respect them. They tell us that giving is good. Doing things that help other people along is good. So go ahead, be nice. Help total strangers, give to charity, spend a little time working for good causes, not because you'll get some cred, or store up treasures in heaven, but because that's what you have evolved to do.[2]

Freeloaders: the way of bacteria

But is there any limit to the power of niceness? Can altruism go too far? Is there a point where you are giving out all the time but not getting anything back? Here's the rub: surely this is a licence for freeloaders. They can let you slave away, typing, phoning, budgeting and reporting, giving your time and efforts freely while they smile encouragingly, and then slip out to grab themselves the copyrights, profits and glory due to you. Eventually you are dumped and they are billionaires. You die in penury and they raise dozens of trust-funded children. Is there anything to stop freeloaders taking over the world?

The lesson comes from bacteria, and it was taught way back. Quite soon after they discovered the joys of stickiness, that they could bond together and attach themselves to the food supply (about four billion years ago), bacteria found there was a downside: making the polysaccharide which forms their glue was energy-consuming. It left them with less energy for making offspring. The colonies' growth was secure, but also slow. Sometimes a colony developed a mutant strain which didn't manufacture so much glue. These could, and did, spend more energy on making babies. They were freeloaders, securing their own success while living off the efforts of others. But as their babies swelled in number another problem arose: the babies couldn't make enough glue either, so the success of the freeloader threatened the cohesion of the colony. Sooner or later the number of freeloaders was so great that the colony came unstuck and was washed away to oblivion. But that's not the end of the story. The vacant space on the rock was taken over by a fresh colony of bacteria, one

28

which hadn't yet evolved freeloaders, and was therefore more sticky. In time that colony would also evolve freeloaders and be swept away, to be replaced by a fresh team.

Later in evolution, the polysaccharide glue was replaced by the social glue of cooperation, binding communities of social animals, be it lions, dolphins, wolves, or vampire bats. It takes time and energy to hunt and gather, and more time and energy to produce and rear young. Those lazy ones who can get off hunting while still sharing in the spoils will have more time to make babies. Their children and grandchildren will end up being in the majority, and because they inherit the lazy ways of their parents, in no time the colony will find itself short of hunters, which leaves it vulnerable, and ultimately doomed. As with bacteria, new, more cooperative and social families fill the niche, until they succumb to the same corruption.

A modern-day equivalent occurs in the world of work. Here is an entrepreneur; we will call him Horatio. Horatio is a man with a plan. He turns it into a small company, works stupid hours day and night, makes a success of it, while sacrificing sleep, peace of mind and his personal life. Like a bacterial colony expands, its profits grow. Eventually they are high enough for him to take on staff; solid, dependable people who work normal hours and allow him to get a life and a girlfriend. The company has reached maturity. As it grows, Horatio takes on more staff. He finds a window in his packed diary when he can marry his girlfriend. Now Horatio is turning normal. He can look forward to children, play, holidays. Among his staff time off can be arranged for illnesses, parental leave, school crises, etc. Life is good.

But out in the market new start-ups are working day and night, sacrificing sleep, peace of mind and personal life to undercut Horatio. When Horatio's sales start to drop, the staff are not laid off or pay reduced; they have mortgages and children after all. Nobody could be called a freeloader, but the rivals are that much more desperate, and willing to cut their margins that much more. When the creditors call in their debts Horatio's company goes into receivership. If only Horatio's staff hadn't all made babies. Like the bacteria colony, Horatio is cast adrift. But as before that isn't the end of the story; the vacant industrial unit will be taken by a new young thrusting entrepreneur, and the story continues, even though Horatio doesn't. Corporate wevolution mimics natural evolution.

If you were ever part of a voluntary organization you must have felt the fierce joy of altruism. The chair of the parent-teacher association of your nearest school is one of those; spending every spare minute preparing for the Bring and Buy sale, furious that nobody else can be bothered to help, wondering if anyone really cares a fig for the school, but knowing they'll all be there on the day, bringing and buying, having a good time, and praising the organizer not a jot; talking to her only to complain at the price of the cakes she spent all week baking. Yet she will do it again next year, and the next. Each year she'll complain about how little help she gets, about how everyone is a freeloader. But she will continue to be the community's glue, holding it together. How many freeloaders can a community support before it collapses? Looking around you now, you might say, 'Quite a lot'.

Being Fair: the way of the capuchin

What to do about the seemingly inevitable rise of freeloading in your community? Can cheats be spotted and dealt with before corruption sets in and your group is swept away? There is another instinct which checks for cheats – our sense of fairness.

In 2003, Sarah Brosnan at the Yerkes National Primate Research Center, Emory University in Atlanta, Georgia, was rewarding capuchin monkeys for giving her pebbles. There were two capuchins in this test, and they were each being rewarded with a piece of cucumber. Then one of them was given a grape instead. The second capuchin noticed this. Grapes are a much tastier treat. She gave in her pebble and held out her hand for a grape, and was given a piece of cucumber. After one look she threw the cucumber back at Sarah. The same thing happened again and again. The capuchin was not just unhappy at not getting a grape like the other one, she started to smash up her cage in protest. The experiment has been repeated on dogs, birds and chimps. On one occasion in the chimp experiments, the deprived chimp kicked up such a fuss at not getting fair shares that the other chimp gave back her grape – she came out on strike in sympathy with her mate. A sense of fairness is something we are born with. 'A fair day's pay for a fair day's work' is a chant recognized throughout the animal kingdom, at least at mammalian level.

Humans everywhere recognize that call too, sometimes without knowing it and often without doing anything about it, openly at least. But they may show their distress in hidden ways. We can winkle out their true opinions by asking statisticians, who can see things that mere mortals cannot. Richard G.

Wilkinson and Kate Pickett decided to measure happiness around the world. On the face of it measuring happiness is not an easy thing to do. You can't just stop someone in the street and say, 'On a scale of one to ten, how happy are you today?' Wilkinson and Pickett looked for other clues, searching among national statistics banks. Generally, if you are happy you don't feel the need to beat up your wife, to steal, rape or murder. You care for your children and those around you and don't use drugs. So for their book, *The Spirit Level*, Wilkinson and Pickett looked for statistics on rates of murder, rape, abuse, theft etc. They looked at educational standards, obesity levels, teenage pregnancies, social mobility and mental health. A wide range of measurable statistics common to twenty countries were sifted with forensic thoroughness. They compared the gap between the richest and the poorest in each nation, regardless of the overall prosperity of the country, and they all pointed the same way: the greater the gap between rich and poor, the unhappier the people. Fairer countries are happier countries. In the UK, the gap between the grapes and the cucumbers is wide and widening. Sooner or later the capuchins will try to break up the cage.

So, while you are giving generously (because you have this instinct for it) there is a fairness meter clicking in a corner of your mind, because you have an instinct for fairness too. And you're not alone. The instinct has evolved all over the planet.

Sharks are known to hover just off coral reefs so that wrasse, the cleaner fish, can nibble parasites from their skin. The cleaner fish would really rather eat a tasty shark steak. Every now and then one of them cannot resist the temptation, and takes a bite. The shark reacts instantly, of course, and chases the cleaner away. But more than that, it won't go back to that cleaner again. More even than that, the other sharks also boycott the cleaner, who will starve for a few days before being let back to work. The sharks punish the cleaner, who will learn not to try eating the sharks.

Even among plants fairness meters are clicking. Leguminous plants such as beans are useful for fixing nitrogen in the soil. But they don't actually do it themselves. They have a symbiotic arrangement with bacteria, who set up home in nodules attached to the roots. The deal is, the bacteria will make nitrogen for the plant and in return the plant will feed and house them. Sometimes the nodules mutate into a form which doesn't fix nitrogen any more. The advantage for the bacteria is pretty obvious: there they sit, being fed regularly by their host, but not having to exert itself so much as to pass the nitrogen back. The plant, though, detects the drop in nitrogen from the cheating nodule, and applies a tourniquet to it, cutting off its food supply so the nodule withers.

In all these cases, action was taken. Unfair behaviour is not just policed, but punished. In simpler, tribal societies such as the !Kung hunter-gatherers of the Kalahari (to pronounce '!Kung' make a click sound before the 'k' sound), cheats and backsliders are rounded on by everyone in a way that leaves them no hiding place. Once they are spotted, a mixture of persuasion, derision and threat of expulsion usually brings them round. If they don't satisfy the expectations of their community, execution is a sanction.[3] This social control can produce strange behaviour on the hunt. A !Kung hunter who bags a big one can expect to be derided by the others. You may say he should be celebrated: he's done something brave and difficult after all, and brought home the bacon. But in the past successful hunters have gone on to throw their weight around. The !Kung know that's a dangerous path to pride. Their mocking reminds him of his place. For his part the hunter performs an elaborate ritual abasement, practically an apology, for having fired his arrow in the wrong direction, which the antelope then accidentally got in the way of. How many managers do you know who can tolerate the mocking of others in their department?

Any manager should have an inkling what is fair and unfair in their dealings with others. If they still think they can get away with a little bullying, humiliating or favouritism, they should know that they are being observed by a power much greater than them, which says nothing but sees everything – the omnipresent instinct for fairness.

Decision-making. The way of the bee

Communication comes into its own when we have to make a decision. There are several ways to reach a decision, generally:

1) Majority vote – over half agree. That means somewhat under half disagree. There will always be dissent in the ranks.

2) Minority rule – by a sub-committee. Takes the decision partly away from the team; a bit of a slap in the face for those left out.

3) Expert decision – an outsider assesses the case and makes your mind up for you. Takes all the pain out of the process – except for the huge arguments about which expert to use.

4) Authority – the boss decides. Allows everyone to claim they've been ignored.

5) Authority with participation – the boss decides after everyone has debated the issues. If the boss can handle this, you're in a very good team.

All of these involve management. Can decisions happen without management? Most managers would say, 'No'. Most animals would say, 'What's a manager?' Epic decisions are regularly made by wild animals who have no training in management at all, migrations, swarmings, relocations, foraging preferences, rubbish clearance, hunting plans. How do animals do it?

Watch bees. When they are swarming, bees are exposed and vulnerable, hanging onto a gatepost somewhere working out where to set up a new home. How do they all come to a decision where to build a new nest? The process is called Quorum Sensing.

Scout bees are flying around the neighbourhood looking for suitable sites. When each one finds somewhere it likes it reports back to the swarm and performs a waggle dance to all the bees in its vicinity.

But how can the swarm bees possibly choose between the different dances of the different scouts? The whole swarm will never get to see all the possible nest sites; not even all the dances. Moreover, the dances themselves aren't truly reliable: a wildly enthusiastic bee will give a much better report than a more lethargic bee, just because it is more enthusiastic. (They're not all exactly the same; there are personality differences even among sister bees.)

The swarm's voting system irons out the differences. If the bees watching a particular waggle dance are impressed, some of them will take off with the scout and have a look for themselves. If the site doesn't impress them so much then when they return their waggle dance will be a more sedate affair, and the enthusiasm in the crowd will fade. If they too are impressed then their dance will be as energetic as the first scout's, the buzz will be stronger, more bees will go to check out the site and their lobbying will affect an even larger section of the swarm. Eventually the buzz for one particular site will be so strong that a tipping point is reached in the swarm as a whole, they have a quorum, and then the swarm has committed; the whole lot ups and offs. The system is good. It wasn't selected by bees, but by evolution over millions of years – and you can't argue with that.

Startlingly, the UK's parliamentary lobbying system is constituted very like a bee swarm. An idea is proposed in a green paper (formation of the swarm). Consultation happens around the country (scouts). Interest groups lobby parliament (waggle dance). Further consultation, committee stage, and a white paper is prepared (more scouts and dances). Eventually the bill is passed into law (the swarm moves to the new nest site). If the British democratic system could be as utterly democratic as the bees', the lobbying system would be as smooth as the bees', however the huge national swarm is too unwieldy, and will never avoid biases, vested interests and the nagging of professional lobbyists.

Organizations should follow the way of the bee when it is time for big changes; moving, downsizing, restructuring etc. Everybody should be part of the decision. Even employees who may ultimately lose out from the changes can be included in the discussions. Their input is valuable, because they are part of the 'family'. If they know all the facts, they become part of the answer. Instead of railing at senior management for losing them their job, they can put themselves in management's position, see it as they see it, and understand what the problems are.

Communicating. The way of the sparrow

Sensing the opinions around you – the 'vibe' – happens in a multitude of ways; through phone calls, tweets, social media or emails. Birds also twitter through the day, not so much through the distinctive, loud song by which we recognise a bird; that only happens during mating and territorial displays. For much of the day sparrows use little tweets called contact calls. The tweets are content-free, but the overall pattern of them gives everyone an idea where the rest of the flock is, as it hops through the undergrowth, so no individual need get left behind. Your tweets are exactly the same, inconsequential chirps which make sure you follow the crowd in whatever opinion is being formed.

The pheromones of bees and ants do the same as tweets, sending signals round and about to indicate what the swarm or nest is up to. For humans it is as important as, but no more important than, any other animal. Pass any school playground at break time and be impressed at the communicating going on there: singing, arguing, shouting, sharing, complaining, laughing and gossiping.

Gossip is a distinctly human chirrup, made available through the unique, rococo complexity of human language. Gossip makes the world go round. You can only pray it goes around the way you would like it to. You will never find out, because by amazing sleight of hand everyone hears dreadful scandals about everyone except themselves. Reputations are built and exploded, fairness is discussed, retribution is plotted, while completely different plots are being laid in the next room. Gossip is always cruel and salacious; nobody discusses people's secret virtues.

Like all things chaotic, gossip is partly good and partly evil. It can aid fairness and control bullies by throwing useful brickbats at the big targets. But it ranges much wider than that. It comments on people's taste in clothes, food and partners. In the past it used to spread preposterous rumours about spying for the enemy, forming pacts with the devil, dabbling in black magic, or just being left-handed ('cack-handed', clumsy or backward). It hasn't stopped either: left-handed, handicapped or dyslexic people are now accepted in our modern, inclusive world, but gingers? Some way to go there. People from the wrong country, wrong religion or wrong side of town all have mutterings to cope with. Much of what we learn as children comes not from our parents, not from our teachers, but from what our best mates gossip about in the playground.

Usually the tabloid tittle-tattle of daily gossip goes nowhere, but if it spreads wide enough, if the stories seem to have substantial evidence behind them, it can lead to action. Our guide here are bacteria, who use quorum sensing when they infect your body. Bacteria are a lot smarter than you think:

A single solitary infectious bacterium sits in your blood stream, surrounded by the cells it is going to infect. It has somehow evaded your body's defences and is ready for lunch.

But it isn't going to show its vicious side yet. If it did, it would alert the immune system and quickly be quashed. Instead it suppresses its lethal nature and hangs around looking innocuous, quietly dividing to become two, then a bit later four, then eight, and so on. Each bacteria cell releases a signal molecule called an autoinducer. It also monitors how many of these there are in the vicinity, so it can tell how many sister bacteria are present. When the bugs have built up numbers to a certain level, the concentration of autoinducers causes them all to simultaneously change to a nastier form, releasing 'virulence factor' molecules that ease their entry into tissues or help them to counter host defences. The bacteria can now launch their attack in such strength that the immune system is swamped.

Gossip works like an infection, with mutterings and rumblings in corners which pass around and grow until there's a big enough crowd repeating the accusation for it to go public. The target of the rumour now has a full-blown attack on their hands, which they have to rebut successfully to save their reputation. Charges of sexual harassment, theft, bullying, embezzlement, lying, moonlighting or profiteering take time to grow, from small whisperings to an open confrontation. In some cases they can threaten the nation: the Arab Spring started small, but grew through classic quorum sensing, tweeting from its beginnings in Tunisia, north-west Africa, in December 2010, right through to Yemen in the East, over several years.

Jimoh, an alpha chimp who was studied by Frans de Waal at Yerkes National Primate Research Center in Atlanta, was furious with one of the other males for mating with one

39

of his favourite females, and began to chase him around the compound. Several females set up a 'woaow' call, which was taken up by more and more of the females. It seems they were not happy with Jimoh's bullying ways, and wanted to protect the other male. Soon all the females were calling, including the alpha female. When it got to a certain level Jimoh realized his life wouldn't be worth living if he carried on (among chimps the phrase 'wouldn't be worth living' can be read literally), so he stopped.[4] The quorum had been reached, and it produced a result.

WHACK!

We see here something opposite to the gentle sweet harmonizing which has been advertised so far. Harmony doesn't work on its own. This is its opposite, antagonism, in full flow. We realize that both are needed: antagonistic harmony creates the ebb and flow of that crazy, chaotic company which is our world. More of this later.

Meetings. The way of the ant

Say what you like about tweets, emails, posts, texts, calls and skypes, nothing will ever replace the old-fashioned meeting. People will fly to the other side of the planet for a one-hour meeting, they're that important.

Meetings follow the 80–20 law: 20 per cent of it is useful. But what is the other 80 per cent for? Many think it is wasted, but there is a lot of ground to be covered in the extra time, because something remarkable is happening. The others round the table are strangers, from another street, if not from another country, yet you aren't trying to eat them. That is one unique difference between humans and other animals. In all other species on Earth an approaching stranger is a potential meal. Ants will drop everything to attack another from a neighbouring nest, then eat it. Bees, termites, lions, meerkats, wolves likewise. So, if you don't want to feel frightened or hungry in the company of the others at the meeting you will need to communicate. There's a lot you don't know about them. They have families, overdrafts and other hobbies which keep them busy for the 16 hours out of 24 which they don't spend in your charming company. If employers want them to remain a loyal and hard-working team for longer, then there needs to be time to get to know one another. They should follow the 80–20 law for the greater good: 20 per cent core business, 80 per cent general talk. A lot comes up in that extra time. The 20 per cent points you to the jobs that need to be done, the 80 per cent makes you want to do them.

Many managers would say that's a sign of laziness. What do ants have to offer on the matter? When I was a child I was told Aesop's famous fable of the grasshopper and the ants, to ensure I worked hard all my life:

The grasshopper played his fiddle all summer long while the ants hunkered down and built up food stores. When winter came, with the snow and ice, the starving grasshopper begged the ants for some food, but they told him he should have been more responsible in the good times and prepared for the bad.

In case you think that everyone should be hard at it all day long, like the industrious ants, instead of lazing around like the hippy grasshoppers, check out some real ants. They seem to be working hard, but actually at any one time a full 50 per cent of them are doing no serious work at all, just mooching. Myrmecologists (ant experts) have done the necessary head counts on many nests to come up with this revelation. That is the good side of scientific research: the discovery that your biggest role model takes time out. The news for managers is that if it's good enough for the busiest creatures on the planet, it's good enough for us. Consider the ant; let meetings take their time.

Although ants look as if they're mooching, you know they are never completely aimless because between them they are creating their ingenious nest. As they drift about they are sniffing all around at the pheromone signals deposited on the ground by the other ants. The pheromones create a rich array of ebbing and flowing colours and currents. When an ant has finished each job and reads the smells around it, it doesn't just react to the first signal that comes along. The final decision on what to do next is taken after a broad survey of the various smells around. If an ant seems to be wandering aimlessly, it's because it is actually trying to prioritise. An ant that couldn't read the pheromone trail really would be aimless. The difference between having no idea what's going on and knowing what's just an inch away is the difference between a nest and a mess.

So, yes, if a meeting needs to meander, let it meander; your team are just finding a new route through the old thicket, looking at problems from all sides, learning how to prioritise. And in the meantime they are giving out vital information about each other, to each other, which helps everyone feel motivated.

Gift giving. The way of the spider

Everyone gives and everyone receives. In the gift-giving culture, even a cup of tea is more than just a bit of hot water, it is the expression of an instinct that dates back to the Cretaceous. You may not like that pink fluffy dinosaur somebody bought you as a birthday present, but the giving of it was an essential act for the giver. So be nice to them. Possibly, when dinosaurs stalked the Earth, they used to give each other pink fluffy mammals.

Why do we do it? Ants offer us a lesson in gift-giving. They present to each other gifts . . . of molecules. Ants don't have a big budget, obviously, but these are the important molecules; pheromones that help all the ants in the area to harmonize.

An ant colony thrives as long as the queen thrives. Her health is signalled by pheromones which she exudes and are rubbed onto the other ants and passed around. If her pheromone signal changes, indicating she is becoming less fertile, the whole colony can read the signal and start to prepare for important changes to come, new queens start to be hatched, and migration is not long away.

Perhaps you think pheromones can't be classified as 'gifts', since they are practical. Yet all gifts have a purpose of some sort. There's no such thing as a free lunch.

Gifts among social animals may convey many complex messages; among less social creatures, when the male gives the female a gift you know he is after only one thing, relating to another important instinct. Male paratrechalea ornata spiders carefully wrap up a present to give to a female. The present need only be the scraps from his last meal; the important part is the wrapping – not the silk it is made from but the pheromone it is infused with. If it does its job properly, while she is unwrapping it, she will allow him to mate. The male Redback spider goes one giant step further and offers himself as a meal to a would-be mate, who spends their nuptials munching on him from the head down. Even the clearly rapturous courtship dance of the Crested Grebe, during which the two lovers gaze adoringly into each other's eyes while performing a perfectly coordinated ballet, ends in a gift from him to her; a beakful of nesting material. Like your boyfriend taking you on an exotic holiday prior to proposing, then giving you not a diamond ring but an ironing board.

Among humans in an office, one very common form of gift, confectionery, can be a bit of a time bomb. Evolution gave us a liking for food (see catastrophilia, page 186), wevolution brought us cookery, mevolution made us compete to see who can bring the most sumptuous yummy to work. Cakes and biscuits are of course bad for you, but not to accept the offer is to reject the giver, which is frowned upon more than a 48-inch waist. If there is a fashion for bringing cakes, then the whole workforce may start to expand slowly but surely until the standard shape is chunky, and it becomes strangely hard for thin people to integrate with the team, because of their odd shape.

Generally gifts are chosen to be neutral. However, if everyone gives you underarm deodorant for Christmas, you might care to have a little ponder, lifestyle-wise.

Displays. The way of the peacock

You may think that dress codes are not a problem for animals, since they are clever enough to be born fully clothed. You must admit when you look at baboons, gulls, rats or peacocks, the effect is fabulous. Well, it doesn't come easy. Birds preen themselves continuously, and furry mammals indulge in lengthy bouts of mutual grooming.

Humans are not expected to groom each other at the workplace, of course. The mind boggles. We must dress ourselves carefully in the morning to avoid having anyone adjust our collar for us. But what to wear? For the bloke there is no problem: suit. Grey suit; that's the business repertoire from top to bottom. There's nothing else possible. The male executive's flat need only be equipped with one tiny wardrobe. Apart, that is, from the space given over to the ties. The only area where a little sliver of individuality can creep in, for eight skimpy inches down the front, is the tie. What significance can you squeeze into that? Your entire ego condensed into an area the size of a Post-it? Many ties must be on display in the wardrobe, so that every morning thirty minutes can be devoted to studying them, meditating on the existential chasm between what statement you would like to make and what you can get away with.

For women it's different. The world of fashion opens out before them every morning. Female executives need wardrobes the size of male executives' flats. For them choosing the right clothes to wear starts well before dawn, beginning with choice of colour, moving on to texture, then material, then pattern, before finally, after hours of agony, they can select the *beret du jour* and move on to the next item. Should we try pastel? Too girly? Or primary colours? Too assertive? Patterned or plain? Austere, punk or new Romantic . . .

One thing in particular has bedevilled women since medieval times – the hemline. A century and a half ago scandals would erupt if a lady showed a hint of ankle. Nowadays the battle has moved upwards to the knee. I happen to think that kneecaps are not a particularly attractive example of evolution's handiwork, yet in certain organizations, women have been sent home for displaying them.

There is one exception; for the IT expert there is a special dress code. He has to wear the opposite of everyone else. He is the only one with a T-shirt, jeans, random hair and three days of stubble. Don't call him scruffy; it's his uniform.

Organizations frequently have Casual Days. The idea is to allow a little of the real world to leak into the office one day a week. On Friday you can wear casual clothes. Here is one company's internal emails in response to Casual Day:

Week 1 – Memo No. 1

Effective this week, the company is adopting Fridays as Casual Day. Employees are free to dress in the casual attire of their choice.

Week 3 – Memo No. 2

Spandex and leather micro-miniskirts are not appropriate attire for Casual Day. Neither are string ties, rodeo belt buckles or moccasins.

Week 6 – Memo No. 3

Casual Day refers to dress only, not attitude. When planning Friday's wardrobe, remember image is a key to our success.

Week 8 – Memo No. 4

A seminar on how to dress for Casual Day will be held at 4 p.m. Friday in the cafeteria. A fashion show will follow. Attendance is mandatory.

Week 9 – Memo No. 5

As an outgrowth of Friday's seminar, a fourteen-member Casual Day Task Force has been appointed to prepare guidelines for proper casual-day dress.

Week 14 – Memo No. 6

The Casual Day Task Force has now completed a thirty-page manual entitled 'Relaxing Dress Without Relaxing Company Standards'. A copy has been distributed to every employee. Please review the chapter 'You Are What You Wear' and consult the 'home casual' versus 'business casual' checklist before leaving for work each Friday. If you have doubts about the appropriateness of an item of clothing, contact your CDTF representative before 7 a.m. on Friday.

Week 18 – Memo No. 7

Our Employee Assistant Plan (EAP) has now been expanded to provide support for psychological counselling for employees who may be having difficulty adjusting to Casual Day.

Week 20 – Memo No. 8

Due to budget cuts in the HR Department we are no longer able to effectively support or manage Casual Day. Casual Day will be discontinued, effective immediately.

Mission statements and mottos. The way of the ant

'Three people were at work on a building site. All were doing the same job, but when each was asked what his job was, the answers varied. "Breaking rocks," replied the first. "Earning a living," answered the second. But the third one answered, "Helping to build a cathedral."' This homily from a speech by Peter Schultz, chief executive of Porsche in the US, provides a helpful introduction to the uses and abuses of mission statements.

Both businesses and ants issue mission statements. Some ant species have empires that cover vast areas, just like some companies. In her super-colony the queen ant may be a long way from the lives of subsidiary nests, and in his multi-national the CEO may be miles from his subsidiaries. Both solve the problem in the same way. An egg is carried from the queen around the provinces, so her pheromone is transmitted to all the ants who touch it. The company CEO distributes monogrammed biros. If ants could make biros they would do that as well. Perhaps the significance of the company-emblazoned ephemera has diluted over the millennia, but its purpose is clear: it ties you to your company. The more you drape yourself in the company flag, the more you sink into the company bosom.

Businesses need them because many of their workers do jobs without an obvious point. It's heartening for someone packing Sony electrical components in a shed in Tamil Nadu that they are 'experiencing the joy of advancing and applying technology for the benefit of the public'. It's very likely they didn't notice that until the email came through from Sony.

No ant has the slightest idea why it is doing what it's doing either. It can't stand back and look admiringly at its work, watch the babies thrive and grow, cheer the other ants along as they return from the fields with the food. The nest as a whole may have a sense of well-being about it, but the individual ants certainly do not. They just follow their noses.

Mission statements tend to be overworded: 'We will be the most highly focused, the most skilled and the most technologically advanced company in the world, providing the highest-quality products and services for customers, delivering consistently good financial results for share-holders, enabling employees to realize their full potential, and ensuring a positive impact on the communities we serve'.

With a little tweaking it would do for ants: 'We aim to be the world leaders in the construction of insect accommodation, organic debris recycling and soil aeration, providing sustainable growth as well as benefits for the local micro-fauna. Through 110 per cent worker empowerment we strive to be environmentally synergetic as well as habitat-competitive, reflecting the dedication of our workforce towards colonizing the world's back gardens'.

No. The mission statement should finish after one line, like the old heraldic mottos: SEMPER PARATUS: 'Always Ready'; OPTIMUM AGE: 'Do your best'; PER ARDUA AD ASTRA: 'Through struggle to the stars'.

Some big companies do have mottos of the right size:

General Electric – 'Progress is our most important product'
L'Oréal – 'Because we're worth it'
American Airlines – 'Something special in the air'
AVIS Rental Cars – 'We're number two, We try harder'
IBM – 'Think'
Canon – 'Image is everything'

Now, those are mottos you could mutter to yourself as you wend your way home in the evening:

'*Progress is our most important product*' as you try to understand the new ticketing system at the station. '*Because I'm worth it*' as you fight to get to the only spare seat on the rush-hour train. '*Something special in the air*' when you end up standing, pressed up against someone's sweaty T-shirt. '*We're number two, We try harder*' as your wife lets you into the house. '*Think*' as you help your son with his homework. '*Image is everything*' as you adjust your hair and pull your tummy in, ready to join your wife in bed.

For many of us, when we are young we have a mission; we want to change the world. A little later we are happy merely to become very rich. Not long after that we discover our ambition is to just keep up the mortgage payments. Our mission statement is like the ant: 'I follow the people around me'.

In conclusion: the Fifth F

The egosystem of the workplace describes the interactions of the people who inhabit it. All are linked by mirror neurons, shared culture and communication. When everyone is generous to others, they are being generous to themselves. Generosity must be balanced by watchfulness to make sure of fair shares. The price of altruism is eternal vigilance.

Zoologists looking at animals often speak of four fundamental instincts which control their life – the 'Four Fs' – Feeding, Fighting, Fleeing and Fucking. To this must be added the fifth – Friending. All the components of friending – joining, giving, accepting, policing, helping, sharing, arguing, celebrating, communicating – all are part of the gregarious instinct which helps us survive and thrive.

CHAPTER 4

THE ART OF SELF-DEFERENCE

Harmonizing can go too far. In order to bond thoroughly with your fellows you may believe it is necessary to suppress any part of you that might put up a resistance. This means doing a whole string of things that you would never be allowed to do by life coaches, business advisers, personal trainers, mums and dads: you learn the art of deference; sacrificing your principles in order to patch in nicely with everyone else. In effect, you give in, roll over and die.

Not all of you dies of course, just the thinking parts. It's surprising how well we can manage without serious use of our brains. As the previous chapter shows, we are remarkably sensitive to those around us. Can't we just let go, let our mirror neurons take over, defer to the opinions of those around and do what they do?

Sure we can. And since the middle of the twentieth century scientists have done a series of experiments, tentative at first, but then with increasing bravado, which demonstrate with embarrassing clarity how far we defer to others. Their results have been both comical and tragic.

Conforming

In 1936, Muzafer Sherif did his famous autokinetic experiment. The 'autokinetic effect' happens when you look at a small dot of light in a dark room. With no point of reference for the light, it seems to float about randomly. Sherif put several people in a blacked-out room and told them the light was going to move about. They were to tell him how far and in what direction. Of course it didn't move, but they still thought it did. To begin they each reported different movements, but over time the people began to agree on the direction and extent of movement of the light. They had each deferred to the opinions of the others, and the result was that they conformed – they all shared the same beliefs.

The experiment showed that in the absence of any other information we not only follow the opinions of others, we believe them to be our own.

In the 1950s, Solomon Asch showed how hard it is to take a minority view, even when you are clearly and measurably right. In a series of tests he asked a roomful of people to say which of three straight lines they thought was longest. An easy enough test under normal circumstances, but this was an experiment, and Asch had loaded it; nearly all in the room were stooges, primed to give the wrong answer on a given cue. Only one person, the 'subject', was being scrutinized by Asch. When the whole room picked the wrong line, it was impossible for the subject to pick the right one in a third of the tests. The subject had never met the others in the room, yet he felt obliged to defer to their – clearly wrong – judgement.

Ants are masters of the art of deference. As they sniff the ground around them they change their behaviour this way or that. No questions are asked, they just go with the flow. It works for the ants – unless there are experimenters in the vicinity. It works for sheep, who tuck their heads between the buttocks of the guys in front and go where they go.* It works for a lot of us humans a lot of the time. The meaning of some of the jobs I have had seemed to wander just as aimlessly as that point of light in Sherif's room. Here I am, typing an agenda for a meeting to discuss the possible content of another meeting some time later. Is anyone going to stick to this agenda, I think. Is the meeting going to reach any serious conclusions? Will it make any difference to anything? Will the world be a better place as a result? If this organization, this meeting, this agenda and I myself were to evaporate now, would anyone out there notice? I realize that provided everyone else thinks the meeting is worthwhile, I will too. Don't ask. Keep your head down and type on.

You can follow the lead of others when they are complete strangers whom you never even meet. In *YES! 50 Secrets From The Science Of Persuasion* (Goldstein, Martin and Cialdini), the authors describe an attempt to encourage conformity in a rather disparate group – hotel guests:

The idea was to raise the number of towels recycled in the guest rooms. Guests are encouraged to help the hotel

---

* Don't knock sheep; they've been around for a good few million years, so they must be doing something right.

save on energy use by reusing their towels instead of having a replacement each night. Notes are placed in all rooms, pointing out the environmental benefits of recycling. For the experiment an additional note placed in certain rooms pointed out that most of the guests did recycle their towels. The guests in these rooms were 26 per cent more likely to recycle than those with the simple environmental message, demonstrating their tendency to conform to what they saw to be the normal behaviour.

Sometimes conformism produces the wrong result. Another experiment by the same team aimed to encourage householders to save energy. Three hundred homes took part in the survey, which involved reading their electricity meters weekly to discover their consumption. A few weeks in, the experimenters started leaving cards at the homes with a printout of the overall average consumption. A few weeks later they found that homes that had been above average in their consumption were reducing their usage – no surprise there. What was surprising was that the homes that had been below average had raised their consumption. In this case the spirit of conformity had quite the wrong effect.

Control

Because thinking is generally harder work than hard work, most of us don't think, we go on automatic, following unconsciously the invisible signals that the others hand down. Chris Frith records, in *Making Up The Mind*, how students in Cambridge in 1968 conspired to influence the behaviour of one of their lecturers without him knowing. He was in the

habit of striding from side to side of the podium during lec-
tures. The students appeared to be all alert when he went to
the left, but as he moved to the right their eyelids drooped
a little, they relaxed their poses, occasionally a pencil would
fall to the floor. When he strode back to the left they perked
up again. By the end of the lecture he had stopped striding
and was spending all his time on the left side. He himself was
unaware that anything unusual had happened.

Keeping strictly legal is difficult when conformism rules.
When one person bends the rules it's bad practice; when
everyone bends the rules it's 'custom and practice'. In many
civil service departments there is an annual allowance of
two weeks paid sick leave on top of the regular holiday,
should you need it. 'Should you need it?' Of course you
need it. Everyone is sick for precisely two working weeks
each year.

In 2009, the British parliament was rocked by a scandal:
nearly all the MPs were found to be fiddling their expenses
to the tune of millions of pounds a year. The scandal was both
predictable and avoidable.

MPs have been playing with their expenses ever since the
expenses system was created 100 years ago. Up until then MPs
didn't need expenses. They were wealthy in their own right,
and used the Houses of Parliament merely as a club, where
they could do deals and plot to spike their enemies. Towards
the end of the nineteenth century, as democracy broadened
its scope and ordinary people began to be elected as MPs
from poorer areas, it was clear that they couldn't function
properly without some kind of subsidy. In particular, they
needed to rent rooms near Westminster while they were in

town away from their homes and constituencies. So a list of allowable expenses was drawn up for the poorer MPs.

This being a democracy, the rich MPs felt entitled to the same expenses, and so the game was on to see how much they could get away with. Since the regulation of expenses was conducted by the MPs themselves, it turned out they could get away with quite a lot.

Then, in 2000, the Freedom of Information Act was passed. Immediately, it was obvious that journalists would be asking for details of MPs expenses, and making mischief with them if they could. Since the law was timed to come into force in 2009, the right honourable members had nine years to clean up their act and hide the evidence. But they did nothing. Not a single MP blinked, because nobody could break rank, not even the most honest – and there were plenty of them – who claimed no more than the basic minimum. The honour of the club meant nobody snitched on their mates. When the storm broke it devastated the government and undermined what little trust the public had ever had in their MPs. The cheats were everywhere; using 'expenses' to redevelop property, evade tax, employ their own family in non-existent jobs, buy extra houses for their ducks . . . The same brilliant teamwork which could rebuild collapsed civilizations was used here to line pockets.

Making a deference

Dynamic conformism often finds a cosy home on a committee. A committee is set up to agree a course of action in some matter, and in many cases management knows precisely what course of action they would like to be agreed. If the commit-tee members are sufficiently wise to this the only problem for the team is to play a suitably convincing charade of consid-ering the alternatives before reaching the desired conclusion. In management's view 'to um–and–err is human; to forego, divine'. The drift comes about by a million small nudges, each a nothing, but all adding up to the inevitable.

The mother of all committees is the government cabinet. Here we find the cream of the nation's decision makers, each promoting the interests of their own department – defence, education, health, business, etc. – while working together to produce joined-up government policies that work for everyone. Yet even the most rational, highly educated, down-to-earth government cabinet can produce the occasional apocalypse when the pressures of conformity are too great. Perhaps the neatest example in recent years was the British cabinet of 1997–2003. Up until 2002, the New Labour government, which swept to power under its charismatic leader Tony Blair, enjoyed unprecedented popularity with its reformist, liberalizing polices. Then it all went weird. The drift towards an invasion of Iraq, which occurred in the UK between April 2002 and March 2003, happened hypnotically slowly, at the speed of a log in a lagoon. Each day a slow, gentle, but irresistible current nudged the nation gently towards an unexpectedly deep plunge hole. At the time the majority of people in Britain thought there would be little point in an invasion. Most thought it was impractical, many reckoned it was immoral and quite a few were concerned it was illegal. But the charismatic Tony Blair urged the invasion because he wanted Saddam Hussein out. Regime change is the worst of reasons for invasion, so for those of us on the outside who foresaw the inevitable disastrous outcome, the fascination lay in watching the succession of prods over the eleven months which caused the noble politicians to opt for the most ignoble act. A surprise convert to the war effort was John Prescott, the deputy prime minister, one of the toughest of political fighters, famous for his outspoken words and firmly held views. John had a history of being a 'man of the people', who spoke common sense. Although he had misgivings in the lead up to the invasion of Iraq, his early belief that there was little reason why we should invade was replaced in the end by his inability to find a good reason why we should not invade, so he let himself gently drift into the plunge hole. If Prescott can reverse his views in order to conform, who can avoid it?

The ghost in the machine

This is the fate of the loyal employees, as the poet says: 'You are like a rock; each day a little more of you is worn down'.

You are ground away by responsibilities, blown by winds of fortune, frozen out by management, flooded with queries, scorched by criticisms, undermined by gossip, chipped at by young rivals and nuked by the boss. The only way you can get through is by being as nice as possible to as many as possible. The delicate art of self-deference makes you, like a pebble in a stream of events, bland and smooth from all angles. No sharp edges. The nicest person in the department.

It comes at a cost, and the cost is You. Prolonged conformity leaves people sapped of any individuality. They lose the ability to complain, for instance. They may find the food in a restaurant is cold and tasteless but they still nod approvingly to the chef as he comes round. When people barge into them in the street they apologize for being in the way. If asked for their opinion they look around the others to see if they can find it. They have difficulty entering rooms, preferring to stand in the doorway.

Chronic sufferers lose the ability to finish sentences. They can't make a decision without it being ratified at a committee meeting. They find, on looking in the mirror, that they always have a smile on their face, even when they are unhappy. They wear a tie at weekends. They can't remember the names of their children.

Having a meeting with a company person has a strange feeling to it, as if they're not really there. They look at you and nod their heads, but there is a ghost in the room – the company, which is silently whispering to them.

At some point over the last forty years they have lost all their free will. This is not altogether healthy, in spite of what their manager might murmur to them while he pats their head, because the company needs people to be able to tell right from wrong, or it will drift down the path to corruption.

But going against the flow is not painless.

To defer or to differ – whistle-blowers

In 2007, Sharmila Chowdhury, a radiology service manager at Ealing Hospital NHS Trust, discovered that two of the doctors in her department were moonlighting, neglecting their duties at Ealing Hospital to do paid work in a private clinic nearby. She alerted senior managers, who acted promptly and decisively, by sacking her. She had assumed incorrectly that 27 years loyal service in the NHS meant she would be respected. Quite the reverse; 27 years service meant she should have learned how things were done and allowed them to continue.

For about seven years up to 2013, Edward Snowden was employed within the US National Security Agency to look for new ways to break into internet and telephone traffic around the world as part of a massive covert surveillance operation. Edward knew his duty was to obey orders and trust his country. But having been brought up in a family both patriotic and moral, it was the higher moral law which compelled him to leak to the *Guardian* the full, vast range of intrusions into personal privacy worldwide. Poor Edward. If only he had no morals, if only he had never read the 4th and 5th Amendments to the US Constitution or article 12 of the Universal Declaration of Human Rights, or the principle declared at Nuremberg in 1945: 'Individuals have international duties which transcend the national obligations of obedience. Therefore individual citizens have the duty to violate domestic laws to prevent crimes against peace and humanity from occurring'. But by retaining his integrity, Snowden incurred the wrath of all the signatories to these lofty principles, so no country welcomed him. He became stateless.

When you are young you are told about moral values; about honesty, decency, integrity. To your surprise, you find that at work these are secondary to loyalty, conformity and obedience. So your morals enter a long dormancy. For so many years you go with the flow, follow the sheep in front, keep your head down and type on. When something happens which kicks you out into the real world, you wake as if from a dream, realize that something is deeply, systemically wrong, and then the real trouble begins. If you talk with your colleagues they tell you not to worry yourself, not to risk your position, not to make waves. The further up you pursue it, the more resistance you get, until you are left with the impression that everyone around you is corrupt, and has been for some time. Further, some of the things you yourself have been doing over the years must have been dubious, edgy or downright crooked. It's pretty traumatic for you personally, but it's horrifying when all your friends turn on you, which is what they must now do.

You can be attacked by your friends even if there is nothing illegal going on, in what is known as the 'Semmelweis Reflex', named after Ignaz Semmelweis, a Hungarian obstetrician who discovered in 1847 that if doctors washed their hands before attending childbirth, the incidence of fatal childbed fever – we now know it as septicaemia – was reduced tenfold in maternity institutions. There was no hint that doctors were corrupt because they hadn't washed their hands up to now; they had never realized there was anything wrong.

What happened to Semmelweis followed four stages of rejection:

His first observations to his colleagues elicited mild interest, but no change in working practices. He was gently smiled at and told to forget about it. But Semmelweis was fuelled by the desire to save lives, and recklessly considered this to be more important than maintaining the status quo. So he campaigned for a change in practices.

In the second stage he received a lot of attention from many eminent doctors. It wasn't the right kind of attention, though. They rounded on him, disputing his claims and his methods and finally ridiculed him. Still he persisted.

Here was the dilemma: by not washing their hands in the past, the doctors had inadvertently been responsible for hundreds of deaths. But by not washing their hands in the future, they might be held responsible for deliberately causing more deaths. There was only one thing they could do.

In the third stage Semmelweis was vilified, his reputation was savaged, his competence called into question, his sanity doubted.

In the fourth stage he was shunned. This was enough to drive him to depression, alcoholism and a nervous breakdown. His wife referred him to a mental hospital in 1865, where he was quickly put in a strait jacket and beaten by the guards. From this beating, in an ironic twist, he contracted septicaemia himself and died. At the same time, but in another part of Europe, Louis Pasteur was formulating the germ theory, which would eventually support Semmelweis's findings. There is now a Semmelweis museum in the house where he was born, and a statue to him in Budapest, but no record of an apology from the doctors.

The shunning of individuals who disagree with the majority is a widespread, even normal pathology. No communication happens, the victim is left out of social activities;

spoken of in the third person, as if they aren't there; there is no eye contact. As if a witch-doctor's bone has been pointed, the victim is as good as dead. Perhaps in the modern Western world the death is merely a 'professional death', but it's real enough in poor countries or in jungle communities, where survival depends on protection, feeding, and grooming against parasites.

Shunning is very nearly the worst kind of punishment. In southern Africa you find the concept of 'ubuntu'. The phrase 'umuntu ngumuntu ngabantu' (a person only exists through other people) summarizes the principles of harmonizing. We only exist because others allow us. Without them we are powerless, friendless, helpless and hopeless.

But the worst kind of punishment is solitary confinement. At this point in the book we come to the farthest extremity from the converging, clustering, flocking, tribal animal which has dominated the world since life began. There are enough histories of the serious psychological harm done by being totally removed from human company for solitary confinement to be one of the 'cruel and unusual' punishments prohibited by the Geneva Convention. This is not logical, if you want to look at such a thing logically. To be separated from one's tormentors should be a blessed release. The fact that it is considered the worst kind of torment shows us the paramount importance of the group instinct in our life.

All our professional lives we are sensitized to the nuances of tone and body language we read in those around us, fearful of transgression. A cross word from the boss, or just the blink of an eye, and we are nudged back into line. The higher up in an organization we get, the more our position depends on the support of our mates, so the more we cannot break faith with them. Even when they seem to be leading us to the brink of homicide we cannot bear to break faith with them.

Armageddon

On the 28 January 1986, the Space Shuttle Challenger disintegrated directly after launch, killing all seven astronauts aboard. The enquiry that followed engaged eleven experts and a maverick. The experts went round in a group, asking the right questions to the right people and getting the expected answers. The maverick, Richard Feynman, refused to follow the proper path, asked quite the wrong questions of the wrong people, and in so doing revealed the degree of groupthink, collusion and mutual myth-making that to be honest, all contractors and agents everywhere tend towards, but which in this particular instance were exposed in a very, very public arena. There was no conspiracy, no greed or self-serving, just the normal sliding away from good practice which happens when everyone practises self-deference.

One example can stand for all: one particular claim, made by a senior manager at NASA at the public hearing, shows how even the most intelligent, highly educated person, under the influence of groupthink, can make statements bordering on the lunatic: he claimed that the chance of a shuttle mission failure was 1 in 100,000. That's an impressive safety claim, which he said was based on the best available statistics.

But as Feynman pointed out, '. . . since one part in 100,000 would imply that one could put a shuttle up each day for 300 years expecting to lose only one . . . we could properly ask, "What is the cause of management's fantastic faith in the machinery?"'

So, is conformity always going to get to us in the end? Are all well-ordered teams going to end up following each other's trails like ants, blind to the bigger picture? Are we doomed? No. There is another animal lurking within us, who is there to do battle with the ant: the ape.

SECTION 2
ANTAGONISM

CHAPTER 5

APE AND ANT

So far we've heard a lot about ants, bacteria, rats, sparrows etc., all held up as mirrors to our own nature. But you may find your pride feeling a little bruised. When we look at the tree of evolution, we are used to seeing ourselves right on the top. Grudgingly we've had to budge along our branch to make room for apes. Now we're being asked to share with all sorts of dubious creatures from lower down, like ants.

Remember, just 150 years ago, before Darwin's *On The Origin Of Species* was published, we considered ourselves to be 'a little lower than the angels' (Psalms 8:5); we were God's gift to the planet – or rather the planet was God's gift to us. Then Darwin kicked us down a bit, and yes, I can see there are similarities between us and apes . . . all right, 98 per cent if you insist on looking at the DNA. But ants? No way is our lifestyle like the ants. We couldn't be more different.

Ants are all equal; we are not

Ants have no managers, generals, authority of any kind. Everyone can have a say about everything. As we saw on page 44 the queen ant is queen only in name. Ed Wilson, the famous ant guru, famously said Karl Marx had had the right idea, just chose the wrong species. Ants are all gene-deep communists.

However, for apes (and for us) some are definitely more equal than others. While ants are all at the same level, we (and apes) are devoted to rank and status. While ants are happy to scurry about messily in all directions, we organize troops, sections and platoons. We obey orders from the top dogs and beat a cadenza out of the underdogs. We like parades where everyone marches in step to the same drumbeat, singing from the same hymn-sheet. We like hierarchies where orders are snapped from the top to the bottom. We want to be shackled with chains of command. To hell with Wikipedia, we love Encyclopaedia Britannica, a glorious human pyramid, with its managers at the top, passing instruction down to heads of department, thence editors, sub-editors and writers.

That's the ape's way, and you must admit it works. How would armies operate if every soldier went independent, each one launching his offensive when he felt like it? How would it be if the publicity department sent out the sales catalogues before the product was designed? Things only work when they're organized, and they can only be organized from one place: by a boss!

Ants are harmonious; we are not

Ants cooperate together, all well tempered, happy to do whatever the other ants tell them via their pheromones (see page 9) whereas we (and apes) disagree all the time. We argue about everything, criticize, debunk and look for the flaws.

We are born to fiddle in this way. As soon as an invention appears we need to change its specifications. As soon as the internet was born, we were designing viruses to test it to destruction. As soon as a piece of legislation is proposed, opposition groups emerge chanting from the woodwork. This is so much part of our life that we have made it part of our institutions: in Parliament, 'Her Majesty's Loyal Opposition' is there to be an official bloody nuisance. And as soon as the House of Commons is satisfied with a proposal, the House of Lords rips into it. At election time, what is it the politicians ask us to vote for? 'Change'.

Hell, from the moment we are born we want things . . . and different things. We want the things our brother has, and we nick them and fight him for them, then we systematically break bits off them (the toys, not the brother), and want new ones (new toys and new brothers).

We continue to fiddle this way when we go to work. You know that feeling when you first get a job with an organization as, say, 'Office Resources Manager'? (Posh-sounding title; you are there to empty out the rubbish bins, actually.) You're young, feisty, you've got ideas. You turn up on your first day and it only takes a minute to see what's wrong with the company. So out of the goodness of your heart you tell them what they need to do to drag themselves into the twenty-first century. But your brilliant scheme is forever being obstructed by the old fogies, the stick-in-the-muds, set in their ways, muttering, 'We don't do things like that here . . . We tried that before and it didn't work . . . That's illegal' and suchlike cowardly nonsense. You realize this will be a long campaign. You go underground. You steadily worm your way into the system. You hang around the corridors of power until you get promoted to a position where you can launch your master plan. And just when you are about to unveil it, a new office resources manager arrives, hired to empty the bins; an upstart teenager who knows nothing and thinks he knows everything, who starts telling you how to run things. The rest of your career is spent trying to keep him down . . . and so it goes on, in a never-ending spiral, young fogies ripping into old fogies, old fogies snarling at young fogies . . .

That doesn't sound like the way of the ant.

Ants are identical; we are not

They are clones, so no wonder they think as one; they *are* one. Humans come in such a variety of sizes and shapes, each with its own little quirks. We pride ourselves on our individuality. Our education system is built around the individual. You are on your own for homework, project work and exams. It's you who gets to university, not you and your mates. You are on your own in the exam room. If in the middle of an exam you ask your neighbour's opinion, you are thrown out. We're stamped through with individuality. We deliberately wevolved it.

Let's sort this out

Section one pointed out how much we are like ants, forever sniffing out everyone else's opinions and following them; now the claim is that we're completely different, rejecting everyone's ideas but our own. We must be one or the other. Which is right?

Let's do an experiment. We will become intelligent designers and create a genome for the perfect human.

Apparatus: We need a bottle of harmony genes, a bottle of antagonism genes and a ton of zombies.

Method: The first thing, says section one, is an instinct that makes us all harmonize like ants. So we pour harmony into all the zombies. What happens?

Results: Everyone throngs into a big cluster, empathizing like mad, sharing, imitating, supporting, praising and adoring each other. It's like a vast, lugubrious love-in.

But nothing's getting done. If one were to suggest something one would appear far too pushy, so our gathering is full of 'Well it's an idea, but only an idea' . . . 'I couldn't possibly' . . . 'After you' . . . 'No, after *you*' . . . 'No, after *you*' . . .

Right, let's swap the 'harmony' for 'antagonism' – a gene which induces criticism, initiative and leadership. Now begins another problem: A's idea is different from B's, so they are going in opposite directions. B can't do what he wants because C is in the way. C has to trample over D. The cluster is splitting asunder. Everyone storms off in a rage.

But what happens if we add back some of the harmony gene? Everyone gets back together and hammers out an agreement. Then antagonism kicks in again and they all dissolve in acrimony. Then they make up, then they start to think independently again and the group splits, then reforms, then re-splits . . .

Conclusion: If we balance these opposing genes nicely, we get life as we know it: continuous debate and compromise, fusion and fission, until everyone is reasonably satisfied. It's a 'social fractal'.

The social fractal

Fractals are a wonder of nature. We didn't know about them until recently because we hadn't evolved big enough brains. Recently we wevolved a brain extension – computers – so now we can start to explore them.

Computer fractals can be built from the collision of two or three mathematical processes, creating fantastic designs which spiral inwards and outwards to infinity. They look like natural forms, and for a good reason – fractals are the basis of nature.

Natural fractals occur all the time when two or three forces work against each other. Fractals lie behind the creation of lungs, mountains, plant leaves and river valleys. Up in the sky, when the forces of air pressure, humidity and temperature interact, you get beautiful clouds. They are fractals too, and follow four basic rules of fractals: a) you know roughly what they will look like, b) you can't predict exactly what they will look like, c) they will be 'self-similar' – their overall shape will be roughly mirrored in the small details, and d) every now and then they will appear to repeat themselves exactly. You can test this idea on a summer's day where the sky is full of clouds: look at them and notice how the shape of small elements is repeated on the larger scale.

An aerial view of the Himalayas shows that the general shape of all the valleys is identical. At least, identical from twenty miles up. Down on the ground individual variations between valleys will ensure you get thoroughly lost in minutes, but because the geology and the climate is the same, the weathering of the mountains has created similar shapes for over a thousand miles.

In our gardens, the beautiful patterns in our plants are all fractals, caused by a genetic instruction about the division of branches and the shapes of leaves interacting with the location and the daily trek of the sun across the sky.

Courtesy of Michael Barnsley

Social fractals emerge among humans from the colliding instincts of the ape and the ant agreeing and disagreeing, flying apart then back together. If we search for self-similarity we will find it. Local politics mirrors national politics. Boundary disputes between neighbours can break out into violence, however trivial the patch of land. Just around the corner a border skirmish could end in the Third World War. Wherever there's a table, there'll be an argument around it: the kitchen table, the committee-room table, the boardroom table, the government cabinet table or the international conference table. The old saying that 'history repeats itself' suggests that human history is a fractal. But the repetition is never exact. Karl Marx came close when he said, 'History repeats itself, first as tragedy, then as farce'.

In my twenties, as a graduate scientist looking for a job in theatre, I had a lot of spare time. I was taken on briefly to make some props for an agitprop theatre company called the Socialist Workers Revolutionary Party Community Theatre Collective. I sat in on one of their rehearsals, carving a piece of foam rubber into a cake (which would later be unequally divided between the fat capitalist and the struggling proletariat). The play was being democratically written by all the actors as they rehearsed. They weren't going to be dictated to by any boss. They had no idea how the piece would run, save that the capitalist would meet some kind of sticky end while the workers sang an as-yet unwritten anthem as they marched towards a glorious future, which I would be painting on a banner after I had finished the cake. The actual scenes and the storyline had yet to be created; plenty of room for antagonism between the thespian egos. The harmony came from the common political cause and the imminence of opening night – they were going to have to agree in the end.

The day I was there they had got as far as the moment when the capitalist villain first entered the stage and did something evil. I never did see what dastardly deed he was to do, because the cast couldn't agree from which side of the stage he should enter; from the left, as the script had it, or from the right, as his political leanings would seem to dictate (because, some pointed out, he was 'right-wing'). After an hour-long democratic discussion it was agreed he should enter from the right. The rest of the morning was a discussion of whether that should be stage right, as the actors would see him, or the audience's right (which is stage left).

We should never forget that the essential basis of fractals, from flowers in the garden, to international trade agreements, is chaotic. (We can include our revolutionaries' final creation alongside Shakespeare on that one level only: Shakespeare's plays were modified by him and the actors throughout his life and fiddled with after he died. Even the Bible is chaotic. God apparently creates Adam twice: Genesis 1:27 and 2:7, and if you want the story of Jesus' life, you can choose the versions of Matthew, Mark, Luke or John. The Good Book was compiled, translated and edited by scholars from 500 BC onwards. Over forty different versions are currently available.)

Are ants and apes social fractals?

Applying the fractal principles to apes and ants, we should find that they are themselves fractals. This is indeed the case. Self-similarity goes all the way down.

Ants are not all harmonious

Ants are a mixture of harmony and antagonism. Watch some and you will see that they spend most of their time in a state of perpetual dissatisfaction, each fiddling with the efforts of the others. An ant emerges from the hole bearing a dead body. Dead bodies have to be disposed of away from the nest, so our ant searches around for a suitable spot and places her load there. As soon as she has left it, another ant picks it up and carts it off somewhere else. And it doesn't sit there for long. Soon another ant is standing over it saying, 'Tsk tsk. Who did that then?', before shifting it again, then another and another heave it this way and that. That dead ant will be carted hither and thither by a string of eager ants, all wanting to do the best and convinced they can do better than the last one. When at last they all stop touching it, when they're all satisfied, then the body will stay there, indeed other corpses will join it, because that is now the graveyard. And since everyone has had their say in its location, it is probably in the right place.

Antagonistic harmony has the same effect on us and our projects. Doesn't that game the ants played with the corpse remind you of the project you designed at work, which was taken by your boss, eviscerated by another committee, the good bits taken out and bogus bits put in, diluted and jumbled until it made no sense any more, then published with your name prominently on the front? All the fiddling took the gloss right off it, but it meant that it ended up where everyone was reasonably happy with it, and therefore nothing will go disastrously wrong, because any blunders have been corrected. More annoying, by the same token, all the bits of genius have been meticulously steamrollered. The result of all the fiddling is something which won't be much good, but won't be too bad either. It won't be the Taj Mahal, but that's alright so long as the plumbing works and the cupboards don't jam. It won't be a masterpiece like Citizen Kane; it might pass for the next episode of your lunchtime soap opera.

Apes are not all antagonistic

When early primatologists studied chimpanzees they thought they saw the human office hierarchy in caricature, with the 'silver-backed male' at the top of the tree. He was big, brutal, and beat up all the males beneath him, who in turn beat up those beneath them. Alpha male, as the books put it, 'had access' to all the females, and the other males were kept away by a variety of rages and displays. Every now and then one of the younger males tried it on with alpha male, but after a lot of screaming around the jungle he would retire to lick his wounds. He then had to do humiliating grooming on alpha to restore his position and stop him chewing his ear off.

That's the classic picture, told by the first researchers in the field, and you can probably guess that it never quite happens like that. Here are some alternative scenes from the lives of apes (chimpanzees in this case) as described by Frans de Waal, Jane Goodall and many of the more recent researchers. You will see they are quite different from the classic tale. Nonetheless they might remind you of an office near you.

1) Alpha male is indeed big and brassy, and throws his weight about everywhere. The junior males are in awe of him. The females, on the other hand, frequently gang up against him and make his life a misery.

2) Alpha male is calm and respectful, generous to the other males and democratic in his dealings, so there is hardly any difference between the top of the heap and the bottom. A species of ape closely related to the chimpanzee, the bonobos, are seriously generous in this way.

3) Alpha male is indecisive, but retains his position at the top by a succession of plots and intrigues that keeps the others so busy warring among themselves that they haven't got time to form alliances against him.

4) There is no alpha male; alpha female sees to that. She and the other females tell the males what will happen and the males obey, or else lay themselves open for a good ragging.

5) In matters of sex, not only does the alpha male *not* get all the girls, evolution would be in difficulty if he did, because the success of any species relies on continuous mixing and melding of the gene pool, which is never going to happen if there's only one source of male sperm. So there are times in a female chimp's life – times which have only been spotted by very assiduous researchers and hundreds of hidden CCTV cameras all around the chimps' habitat – when the females slip behind a tree with one of the other males. Evolution expects it; gene diversity needs it; and lo, the species provides it. Studies of chimp DNA show that in many cases the chimp who is alpha in terms of being loudest, strongest and most dominant is not most dominant when it comes to spreading his DNA.

Furthermore, researchers note that arguments between chimps, though vicious at the time, are nearly always followed by reconciliation and mutual grooming, started by the winner.

Simple hierarchies don't always happen in the office either. On first glance at a team there will often be one who is loudest, pushiest and seems to be in charge. But often they are being merely tolerated by the others, who wait until they have gone out before making the real decisions. In the average department or team, everyone is an alpha in their special zone. One will be an IT expert, another may be better at chairing meetings, another at preparing spread-sheets, another at understanding the pathology of photo-copiers. The would-be alpha boss has to humble himself before Photocopy Queen, who must bow before the mighty geek when it comes to converting a spread-sheet into an interactive bar-chart.

NYAHAHAHAHA!

Sometimes there's too much harmony, and antagonism needs to be injected. In the early days of the Christian Church the ultra-harmonious bishops would meet to interpret the mysteries. Knowing how easy it would be for a collection of like-minded people to agree without knowing why they agreed, they introduced a 'devil's advocate' – one of the group would pretend to disagree, and attempt to ridicule the others through argumentation. (In the absence of any scientific knowledge, all 'truths' were based on argument. The early universities didn't teach science, just grammar, logic and rhetoric.) The devil's advocate is still with us, alive, well and sitting on all self-respecting committees.

Arguments are an essential part of our life. How eerily quiet the world would be without them. Something would seem wrong, like walking past a school playground without being deafened by the screaming of the kids (all arguing, of course).

So beware of anything which hasn't been argued over. There must be debate. It must never go through on the nod. The greatest threat to good governance is the rubber stamp.

The Social Fractal in action: The Starling Fractal

An example of natural, organized chaos happens daily, very close to where I live in Brighton. Every evening a huge flock of starlings wheels around the piers in a spectacular display of coordinated flying. The shape of the flock changes, flattens, bobbles, extends arms, reforms into a ball, for half an hour before they all roost for the night. Where do the shapes emerge from? Who is giving the commands? Where's the alpha starling? For many years the flocking mystery was impossible to solve – it was just too big and complex. But with the rise of the modern computer the problem came within reach of scientists. The discovery they made was surprising. It turns out that nobody is giving any commands, and nobody needs to. The birds all obey two rules: 1) get to the middle, 2) avoid bumping into each other. The voluptuous shape which emerges is the sum of all the starlings' simple rule-following. In 1986 Craig Reynolds' computer animation 'Boids' appeared. His boids were each programmed with those simple rules and behaved exactly like flocking starlings.(And they've been flocking ever since, on www.red3d.com/cwr/boids.) So, spinning around the pier, a chaotic and unpredictable pattern emerges from two simple rules, followed by everyone. If someone did try to organize the starlings, the flock couldn't happen. (What made this difficult to simulate on older computers was not programming in the rules themselves, which are simple, but the power needed to repeat the rule for thousands of simulated birds and the millions of interactions that happened between them.)

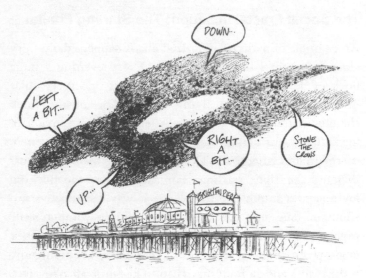

. . . and the Human one

Humans do something similar. Every evening, while the starlings are converging around the piers, flocks of commuters teem through the main concourse of London's Victoria station. They too follow a simple set of three rules: keep to the middle of your particular 'flock', keep your distance from

the others and, in my case, head for platform 17, that's how to get to Brighton and roost for the night. Anyone who breaks out from following their rules – sets off suddenly in another direction, or stops suddenly, will bump into someone else and mess the whole thing up. Humans seem to follow rules similar to the starling – in the evening, at least. Yet there is no organization. Our flocking behaviour is the result of simple laws which have an astonishing effect when viewed from a distance. Victoria station is chaotic, but it works. An alien zoologist, looking down from his flying saucer, would marvel at the complexity of life in railway stations. He would also look in vain for a supreme organizer.

Among other animals we can easily find more examples of apparently, but not, organized behaviour. The whole world is at it. On the plains of Africa, mass migrations of wildebeest begin without a siren sounding or timetables being issued. Elsewhere locusts swarm, salmon spawn, buffalo roam. In all cases we search for the management who authorized and scheduled it; in all cases we don't find it.

This is all elementary stuff. Those flocks, herds, swarms and commuters are pursuing simple, short-term goals, so maybe just following a few simple rules can work for them. But looking at the bigger picture, human civilization seems to be a much more splendid thing. Surely we couldn't create all the diversity and richness of our trades, industries and culture without more diverse and rich organizational structures. Are there any animals that build empires like ours, but chaotically?

The Ant Social Fractal

Ants are the ultimate empire builders. They have yomped around the world since they first appeared 100 million years ago, unstoppable. If you are looking for cooperative behaviour, this is definitely the place to look. And as you peer closer at the nest you will find more and more that they remind you of us.

Ants specialize, as we do. Some handle the brood, some forage, some build and repair the nest. They change roles as they grow older and more experienced. Depending on circumstances they may swap jobs – when they are short of food there are more foragers, if they are invaded they all run to defend the nest and protect their young. We do that. They can become farmers like us, creating fungus farms or milking domesticated aphids for their honeydew. Ant nests last about fifteen years on average – the length of the queen ant's life – and during that time the nest goes through stages of development, from frisky youth, through sedate maturity to dysfunctional old age. Companies last thirteen years on average,[1] and they go through the same stages.* The similarities add up.

Here's a difference, though: Look closer at the ants' nest. There's no management. You would expect to find the queen ant ensconced on a throne, issuing edicts. But the queen is two metres below ground, busy laying eggs. Laying eggs is all she does for all her fifteen years. So the nest is left to fend for itself; a formula for disaster, you might think.

Zoom in more closely; focus on an individual ant. It has a familiar look about it. It's you in miniature! Bustling around, saying hi to everyone it passes, picking something up, putting it down, fiddling about, saying hi to all the new passers-by, following them a bit, then going back to the thing it picked up and put down, finding someone else has picked it up and is putting it down somewhere else. A life so pointless, you might think (and so would they if they had brains big enough to do thinking), yet each ant is actually important. They are each doing what seems the right thing at the time, and what turns out, if you step back and look again from a distance, to actually be the right thing. Because the nest works. It works even though each individual has no idea what is happening more than an inch in front of its nose. They can't guess what

---

* Ellen de Rooij of the Stratix Group in Amsterdam indicates that the average life expectancy of all firms, regardless of size, measured in Japan and much of Europe, is only 12.5 years. A very few companies have lasted for centuries – The Sumitomo Group has its origins in a copper-casting shop founded by Riemon Soga in the year 1590. Most companies keel over well before their tenth birthday. Ant colonies, likewise, are inclined to go extinct in a couple of years. Some last 30 years, but most of those studied and protected from disaster have 10–20 years of life in them.

the overall plan of the nest is. They don't even know there *is* a nest. Ants are stupid. Life is just one thing after another, and somehow it all hangs together.

Remember this as you shuffle home from your monotonous job, shrouded in nihilistic fug, convinced of your pointlessness in a pointless world. Don't get hung up about it, because while you are getting all depressed, the ants will take over the world.

NHS

For a fine example of a social fractal I give you the National Health Service (NHS) in the UK; an institution which is seldom out of the news. And why? Because it's a mess. But the ant would say, don't look too closely. Remember, an ants' nest looks a mess when you study it close up. Observe any single ant and the sense of purposelessness is pathetic to behold. It blunders about this way and that, seemingly at random. Deborah Gordon of Stanford University, who studied them for decades, said that they seem so useless it's hard not to stoop down and give them a helping hand every now and then. Yet the result of the chaos is their magnificent nest, a success story that has lasted 200 million years.

Look at the NHS and you will see the same. Every day the papers tell stories of the awfulness of it: overspending here, understaffing there, underspending there, overstaffing here. Chaos! But if you are observing from another country, or if you receive life-saving treatment for free on the NHS, you think it's a marvel. The NHS is the envy of the planet.

The chaos of the NHS is churned up by the battle between ants and apes. The ant is caring, full of empathy and generosity – exactly what we need in our distress, while the ape is watching the pennies and keeping the records – vital to the health of the organization. If left in sole charge, the most brutally efficient apes would cut costs to the bone and have a patient in hospital for no longer than it takes to poke something in or saw something off. No cash wasted on preventative care, or aftercare, or any care at all if it took time. On the other hand, if it was given over to the most wonderfully perfect, loving, giving, caring, sharing ants, it would be bankrupt within days through financial neglect. You can't have one without the other, and this is true of all enterprise.

So close up, life's a mess; from afar it works. What about your job? A meaningless routine, full of lost opportunities, humiliation, frustration, days bulging with bureaucracy, bullying and boredom? Look at it through the history books and it adds up to Western civilization, which ain't bad, on the whole.

On any timescale you look at, it's a changing world. Business changes, technology, the political landscape, even the planet itself changes as drought, disease, fire and flood make a mockery of our careful plans. So if you are daft enough to build a rigid edifice of rules and procedures in a world that constantly shifts around, it will crack. You should swap foundations of principle and pillars of legislature, which are too brittle, for floating rafts of principle and soft beds of legislature, which can flex. Don't be afraid of chaos; be prepared for it.

CHAPTER 6

ALL APE

We have seen the uncomfortable result of too much compliance – pure ant with no ape. What if the ape is unleashed without any ant? This happens all too often, because changes in the egosystem have forced the ant close to extinction, while apes now stalk the Earth.

The Descent of Manager

From the moment Thomas Newcomen unveiled the first steam engine in 1710, the pistons of the Industrial Revolution have pumped faster and faster. Our ancestors swarmed like ants towards centres of industry such as Manchester, where they built factories, hired workers, then looked for managers to administrate them.

In the days of yore, out in the farming villages, the old-fashioned way of finding a leader was good, but slow. The whole community would watch the younger generation as they grew through childhood, jostling and arguing their way around the fields and cottages. By the time new leaders were needed, everyone knew who they were – they had emerged over the previous twenty years. This was an ant way of doing things – everyone had a say, so everyone was reasonably happy with the leaders they elected, notwithstanding they would step in front of them now and then to set them right about this or that.

This was fine in slower times, but now businesses erupt suddenly, full of departments that need manning immediately, and those teams need managing, and those managers need special understanding of exotic things – economics, national and international trade law, health and safety rules, etc. The old system will not do. What is needed urgently is instant, off-the-peg managers. Colleges have stepped forward to provide courses in business management most commonly leading to an MBA (Master in Business Administration). On an MBA course the student is trying to absorb in a couple of years what experienced managers take decades to understand, and then some. The young turks work hard. There are many facts to be remembered, many lessons to be learned, and not many weeks to squeeze it in. Nevertheless they can become effective executives, who can be posted to pretty much any department in any industry and help it meet targets and deadlines.

Unfortunately, this forced breeding programme sometimes creates mutant strains.

Unnatural Selection

The problem with MBA courses is that they tend to attract an unusually large number of a particular type of person. Every vocation has this tendency. People in tune with their emotions work in theatre; authoritarians enjoy working in the prison services; the Army attracts a disproportionate number of sadists. MBA courses seem to fascinate people with a blend of these qualities – emotionally unstable, sadistic megalomaniacs. In their hands the qualification turns from a useful tool into a weapon of mass destruction. There may not be many of these mutants, but they are all memorable. Meet them once and you are irradiated. They're so toxic, you can receive a lifetime's exposure in an afternoon. You will suffer from the fallout for the rest of your life – nightmares, cold sweats, that kind of thing. I stress that in the right hands an MBA is an invaluable tool, but in the wrong hands it produces Ape-Manager.

Out in the rainforests, primatologists sometimes encounter rogue males who have been thrown out of their own tribe by the concerted actions of all the males and females, fed up with the trouble they cause. They wander the jungle until they can find another tribe which they can successfully usurp. In the jungle they have difficulty surviving, but in the modern business jungle there seem to be many companies who will willingly take rogue males in.

We need managers. But we never ever need Ape-Managers. A responsible manager understands his team's desire to harmonize and uses that as a motivator; Ape-Manager just issues laws. A careful manager encourages their willingness to please; Ape-Manager exploits it. A respectful manager gives credit; Ape-Manager takes it. A good manager earns respect; Ape-Manager demands it. A generous manager gives gifts; Ape-Manager offers bribes. Meetings are an opportunity to explore relationships and share ideas for a sociable manager but a waste of Ape-Manager's time.

Ape-Manager has his own agenda. His ultimate goal is power for power's sake, and he will do anything to achieve that – plot, lie, wheedle, spy, cheat – he'll even tell the truth if it helps advance his career. He is in touch with his inner ape, and *his* inner ape is an alpha's alpha. With an MBA to wield instead of a branch, Ape-Manager can cut a swathe through any company he joins. What makes him doubly dangerous is: a) he is trusted by the company that takes him on, and b) he trusts himself.

Ape-Managers are sufficiently common to have attracted the interest of psychiatrists. Robert Hare is a criminal psychiatrist who spent a career in prisons studying pathological liars, master con artists, and heartless manipulators who had inadvertently murdered their victims. His Psychopathic Checklist is the standard tool for making clinical diagnoses of psychopaths. In 2002 he gave a talk in Newfoundland to the Canadian police in which, after a display of the usual slides of criminal psychopaths, his startled audience found themselves looking at Bernard Ebbers of WorldCom and Andrew Fastow of Enron. According to Hare, these, and many other modern-day CEOs, showed every symptom of being psychopaths.

Hare pointed out that corporate bosses score high on eight traits common to psychopaths: glibness and superficial charm; grandiose sense of self-worth; pathological lying; conning and manipulativeness; lack of remorse or guilt; shallow affect (i.e., a coldness covered up by dramatic emotional displays that are actually playacting); callousness and lack of empathy; and the failure to accept responsibility for one's own actions. Does that sound like anyone you know?

There may be a reason why psychopaths are found head-ing up top corporations: the corporations are themselves psychopathic. Their sole motive is to improve shareholder value. If they were to deliberately do anything that lowered profitability, their shareholders would sell their stock and the corporation would quickly wither. So they must enslave as many as possible, pay as little as possible, and lie with the best, glibbest charm possible. Thus cigarette companies must defend smoking, energy companies must deny man-made climate change, fast-food chains must tout their obesity-inducing products as 'healthy' and the armaments industry must always be selling 'defence' weapons. All corporations must make everything as cheaply as possible, regardless of the human toll or environmental cost. All this requires heartlessness, lack of empathy and plenty of lying. Not surprising that Ape-Managers stalk the Earth; psychopaths of a feather flock together.

I worked with one company as a sub-contractor, so I was able to observe a psychopath at work from the sidelines. It may have been a couple of decades ago, but I still wake up screaming. In the story that follows we see the swathe of destruction caused by Ape-Manager. Not your normal MBA, for whom being a Master of Business Management is a badge of honour and an asset to their company, but those for whom 'MBA' stands for 'Mediocre But Arrogant'. The names have been changed to protect the innocent. And also to protect me from the revenge of the guilty.*

---

* We will also call him a 'he' for simplicity, with apologies to the entrepreneurial 'she's.

The story

Somewhere in the middle of England a small family firm, Luvly Lady Lingerie, was slipping slowly beneath the commercial waves. It dealt in mail-order fashion lingerie, and for thirty years it sent reasonably adequate catalogues to a list of long-established clients. However, the clients were now dying of old age, they were that long-established, and the company's future looked gloomy, unless they diversified into funeral shrouds.

The founder was ready to flounder but his son wanted to take control of his inheritance and revive its fortunes – that is, make money out of it. He knew very little about commerce – he had got himself a degree, but it was in media studies. He decided to hire an expert. When Moriarty answered the advert he seemed the ideal choice. He was ambitious, dynamic, confident, about the same age as the son. Most especially he was an MBA. The young heir was overawed by the alpha-ness of Moriarty and slightly dazed by the sophisticated business terms that Moriarty used: 'Uplevelling to a new demographic profile, leveraging the talent base and reducing the time-to-market . . .'

Big, savvy words (translation: 'Selling to rich people, finding someone in the print room to design it cheaply and print it quickly). It looked as if they were heading for the big time. The staff were slightly anxious about the new turn of events. And rightly so. Moriarty was not interested in their lingerie, their livelihoods, and certainly not their persons. He was set for a brilliant career in business, and this was going to be the first triumph in his portfolio.

RICHARD ROBINSON

The events that followed went through the six classic stages of a campaign:

1) WILD OPTIMISM

2) TOTAL CHAOS

3) UTTER DESPAIR

4) THE SEARCH FOR THE GUILTY

5) THE PUNISHMENT OF THE INNOCENT

6) THE REWARD AND PROMOTION OF THE INCOMPETENT

Stage 1) Wild Optimism
Actually the era of optimism was a short one, owing to the early onset of chaos. With most enterprises there is an eagerness to celebrate the triumphs to come nice and early, before reality breaks in. Moriarty pre-empted that. He was a man with a mission. So he told them the plan he had devised straight away: to target the granddaughters of the 'long-established clients' with a new line in more provocative lingerie. A master stroke! The old biddies would now receive through the door a magazine with scantily clad, lush-limbed lovelies on the cover, arrestingly lit in their skimpy underwear – less like the dowdy catalogue of the past, more 'classy'. Naturally the biddies would be delighted, and would without hesitation give it to their darling granddaughters when they popped over.

Moriarty outlined the project in fifteen meticulously detailed stages. Confusion broke out immediately. Fifteen stages was ten more than they could take in at one gulp.

98

At the second meeting Moriarty had cut the tasks down to five. The other ten would be slotted in at quiet moments between now and the end of the project in three months. More confusion immediately. Three months? Is that all the time they had?

At the third meeting Moriarty reassured them that although three months was all the time they had, he had been able to secure a budget increase to pay for the extra costs, so now they had £10,000. More confusion and disbelief. Only £10,000? For a project this size?

Some of the older workers began to raise objections. Did Moriarty consider that the clients might confuse the catalogue with pornography? They were cut off abruptly by Moriarty with a brief quip: 'If you're not part of the solution, you're part of the problem.'

Mark Twain said: 'The man who is a pessimist before forty-five knows too much; if he is an optimist after it he knows too little.' The younger ones' enthusiasm was unconditional because they were too young to know better, the older ones' enthusiasm was mechanical, because they were too old to care, but they cast a wary eye over the budget, timescale and workload, factored in their view of Moriarty's maturity and formed their own secret opinion. If they mentioned any doubts or quibbles openly they got a variety of answers, relating to the nobility of loyalty, the thrill of adventure, the stigma of cowardice, the anticipation of success, the ignominy of failure, and ultimately the threat of a summary sacking. Moriarty had upped anchor and thrown away the lifeboats.

Soon enough some of them left for gentler climes. The most likely ones to leave were the ones who were good at their jobs. They could find work elsewhere, while the ones who were left couldn't do that, as Moriarty gradually discovered.

Leadership

Moriarty wanted to get the team motivated as soon as possible. This needed leadership, which Moriarty was looking forward to giving.

Problem was, Moriarty was not actually very good at leading. This may come as a surprise. You might think that was the course in which he just qualified. But MBA courses don't trouble much with people-handling skills. Seven-eighths of management courses are not about people, but about economics, globalization, legalities, competition, diversification, ownership, property rights, financial data interpretation, health and safety, positioning strategies, marketing strategies, stakeholder interests, sustainability, corporate ethics, copyright law, financial management, investment, entrepreneurship, innovation, technology, international marketing, corporate social responsibility and employment law.

The personnel side of the course was covered by such insightful diagrams as this:

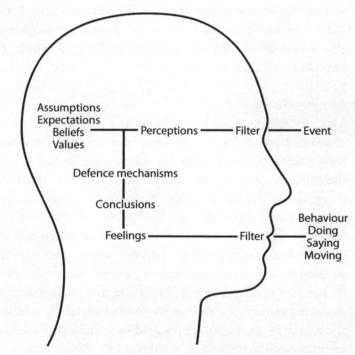

motivation = expectations that work will lead to performance
× expectation that performance will lead to reward
× value of reward

Art (from MBA manual) of simplified psychology

In fact the only actual humans he had come across were the other ones on his course. It's easy to get the idea that everyone else on the planet is just like your classmates, but that would be a dreadful mistake, particularly those on MBA courses. Would-be managers are motivated, ambitious and committed to the ape story. Their employees on the other hand, are more modest. They're in it for a regular wage, which will help their other life; families and the like, something MBAs don't have. They aren't devoted to the job, though they can be devoted to their colleagues. They love good company, not *The* Company.

80 / 20 rule

Moriarty felt there needed to be a more heroic atmosphere – more drive and commitment. Prior to his arrival, the team maintained a steady flow of work which followed the 80/20 rule – 80 per cent of the time was spent on meetings, committees, strategic discussions etc. – and 20 per cent on actual productive work. Moriarty spotted that this was only 20 per cent good for the balance sheets and 80 per cent bad. His task was to raise the output level from 20 per cent to 80 per cent, and reduce peripheral activities from 80 per cent to 20 per cent. He imposed new work practices, enforcing punctuality, shortening tea breaks and lunchtime, and giving them targets, encouraged by various punishments and shouts.

The blame culture had been installed. Everyone now needed to know exactly who was responsible for what, so they would know whose fault it was. They brought in rules and procedures, copied each other in on all emails.

Before long 80 per cent of the time was spent on internal processing and compliance and 20 per cent on actual productive work.

He found his team were not as motivated as they ought to be. They kept on trying to get home to their families occasionally, to eat meals, sometimes to get to the toilet. He started to time them when they went to the toilet. There's nothing that inclines you more to want to spend hours in the loo than that.

Stage 2) Total Chaos

Friendly Competition
Moriarty gave them individual pep talks, encouraging them to be more ambitious. He also gave them group pep talks about teamwork and cooperation (he didn't consider the clash between ambition and teamwork). Teamwork means everyone helping each other out and going at the same pace; ambition means beating everyone else. One descends from the ant, the other is inherited from the ape. The two don't sit easily together. There's no such thing as friendly competition.

Chinese Whispers
Moriarty had minimal actual marketing experience, but he reckoned that was an asset. One should never let facts get in the way of a good vision.

The others couldn't see his vision. But that's all right. Moriarty would feed the vision to the toilers bit by bit, and they would apply their expertise to make it happen. The team were dutiful, but step-by-step management is the Chinese whispers of the business world. To do a job well you need to know not just what to do but also why to do it.

Proper managers have established the goal with their team beforehand, so that they all have an idea where they're going, but Moriarty cut that bit out to save time. His team were in the dark, stopping at the end of each task he set because they had no idea where it should be leading, or perhaps because they feared where it would be leading. If they knew more they could have worked together to iron out little difficulties, but that was not to be. This was Moriarty's toy and no one else was going to play with it. All the inbuilt ability to cooperate and empathize were being squashed.

Motivational Speeches

Moriarty's speeches became full of exotic metaphors:

It's no good staring up the steps. We must step up the stairs.

What the mind can conceive, the mind can achieve!

The difference between the ordinary and the extraordinary is that little extra.

It's not the size of the dog in the fight, it's the size of the fight in the dog.

Pressure makes diamonds.

The difference between try and triumph is a little oomph.

If you fail to plan, you plan to fail.

If you're not fired with enthusiasm, you will be fired . . . with enthusiasm

The team's job was to be bobble-heads, performing synchronized nodding at Moriarty's every suggestion. The result was captainitis.

Captainitis

You have probably found yourself obeying people in peaked caps simply because they have a peaked cap. This is captainitis. The peaked cap, mind you, is not so much the uniform of the captains of industry as the field marshals of the front desk, where it is known as 'commissionaire's disease'. You believe they must be an authority, or else they wouldn't have earned the peaked cap. They think the same, for the same reason, and adopt a special lofty attitude for you. The more you are impressed by their hat, the more they impress their hat upon you.

If a manager catches captainitis, the onset of symptoms is rapid. Moriarty thought he was perfect. Nobody dared question him, so he thought they thought he was perfect, so he reckoned it must be true.

Signs of this perfection:

He called a meeting while everyone was out on jobs. Obviously, he said, it was their fault for being out, not his for calling the meeting at the wrong time. ('Quite so,' they agreed.)

He called for a 5,000-word report, then complained it was too wordy, it should have been 500 words. ('Anyone could see that,' they nodded.)

One day, in a display of hubris, he took on the photocopier single-handed, when it stopped working. He strode over, pushed everyone out of the way, opened the front and rummaged about inside until something snapped off and the machine was ruined. ('It needed replacing anyway,' said they.)

Dissent

There was one of Moriarty's minions who didn't join the rest; who didn't kowtow so regularly. She should have left early on, but remained behind out of loyalty to the founder. She presented Moriarty with carefully researched arguments to the effect that the campaign was flawed, and what should be done. Moriarty deployed bullying.

Bullying is very common in the workplace, and the evolutionary reason is pretty easy to find out there among the other animals. Pushing yourself to the front and pushing others to the back is a fact of strife. Rutting deer do it. Chimps, baboons, dingoes, hyenas, Komodo dragons, vultures, crocodiles and boa constrictors do it, as you might expect. But so do pretty pandas, cuddly koalas, big fluffy polar bears, comical three-toed sloths, colourful clownfish

– they all bully, to our shocked surprise. Probably amoebas do it, very slowly. Why it hurts in the workplace is that there is no escape. Once the bully has pounced, the correct procedure according to the laws of nature is to retreat and keep retreated. The correct procedure according to the laws of the company is to stay put and consider discussing it with HR. Human Resources will treat you like a human resource: expendable and replaceable. They'll tell you your legal position. They won't counsel you beyond offering you a consolation sweet. If you are seriously stressed they will recommend a psychiatrist to help you pull yourself together. So the victim cannot escape. Like the poor wrestler in the ring who keeps rebounding off the ropes straight back into the arms of Zoltan from Bolton, the victim is banged up with the bully. Bullying will recur unless a good ganging-up against the bully can be arranged by others in the workplace, which means much screaming around, if chimpanzees are to be the model (see page 81), but there are serious risks of retribution if you do so with humans.

Stage 3) Utter Despair

Body Language

A primatologist would know, from his study of apes, what the team's body language was saying – despondency, fear, confusion; it all showed in the lack of eye contact, monotony of voice, absence of fun, stiffness of posture. These were warning signals which Moriarty should have been reading. But he was becoming a little bit stressed. You couldn't say anti-visionary things like 'that won't work', or 'I've had a lot of experience in this field and I can assure you . . .' or 'have you considered changing your medication?' A questioning comment was seen as a threat. Worrying was a sign of weakness. Only submissive behaviour came through as support. The toilers learned not to question, not to worry and to do plenty of submitting. Moriarty said he wanted aggressive sales, but only allowed submissive people. He wanted them to show initiative, but not at the expense of obedience.

Hiding

When Moriarty needed something done he would look around to see who was free. Everyone would be preoccupied, using a variety of techniques to do the human equivalent of hiding among the trees; staring at the computer screen with a worried frown, even when it was the screen saver they were scrutinizing; phoning someone's answering machine and holding a long conversation with it; only carrying coffee if they were walking rapidly and purposefully, as if to an important half-completed job, always carrying some papers, never carrying newspapers, never ever carrying newspapers and a cup of coffee.

Friendship

Suddenly Moriarty found that nobody was doing anything. The louder he shouted, the more they huddled. He had tried pep talks and bullying, but the ungrateful, stupid workers were seemingly happy to commit corporate suicide. Morons. There was nothing for it – now he turned to an untouched section of his MBA manual: 'Friendship'.

When the bully cuddles up to you, that chills the blood.

You remember the saying 'a pat on the back is only a few inches from a kick up the bum'?

Moriarty started smiling at everyone. The smile served to increase stress levels dramatically, due to the psychological chasm between the act and the intention.

He promoted everyone, or rather retitled them, so one day the receptionist was re-christened 'director of first impressions'. The girl in charge of the photocopier was suddenly 'reprographics manager'. Who were these new people: 'washroom operative', 'ambient replenishment controller' and 'regional head of services, infrastructure and procurement'? As it turned out, lavatory cleaner, shelf stacker and caretaker.

He instigated a reward system for good performance: gold, silver and bronze stars. If it works for primary school children, it'll work for office workers, he reckoned.

Survival Of The Fitments

At this point the inanimate world showed a hitherto undetected malicious streak. Things started to go missing. Moriarty put his Filofax down and seconds later it wasn't there. Later on it turned up exactly where he had left it. Suspicion fell on certain of the staff in accounts, who had moustaches. Were they a terrorist cell? They were discouraged from entering the office. Tea breaks and lunch breaks were cancelled as a counter-terrorist measure. Everyone began watching everyone else.

Murphy's Laws are familiar to us all as little hitches that bug us regularly. In a troubled office Murphy's Laws begin to rain down, and this was no exception. The computer would crash one microsecond before the 'save' button was pressed.

The vicious but hilarious email about Moriarty which was sent to a few trusted intimates in the office was somehow carbon cc'ed to Moriarty. Broken things mended themselves just as the repair man showed up, then broke again a minute after he left. The lift was always stuck on another floor when anyone went out to post a letter; the letter remained on the desk. Everyone they phoned was either in a meeting or just about to go into one.

It seemed as though the space–time continuum had shrivelled up in despair. Everyone put in longer hours, but the clock monster took them out again. Our perception of time is famously faulty. Time rushes by when we're having fun. It drags by when we're waiting for the photocopier to come free. For Moriarty's workers, ticking off the seconds before they could go home day after day, time stood still.

Murphy's Laws

At the end of the day, there are only a certain number of hours in the day. Not that we wouldn't like to manufacture a few more if we could. Many of Murphy's Laws are based around time and the need for more of it:

Time Law I : The best way to be late is to give yourself plenty of time.

Time Law II: If it wasn't for the last minute, nothing would get done.

Time Law III: The only person who got things done by Friday was Robinson Crusoe.

Time Law IV: The first 90 per cent of a project takes the first 90 per cent of the time, the last 10 per cent takes the other 90 per cent of the time.

Hofstadter's Law: It always takes longer than you think, even when you take into account Hofstadter's Law.

More's Law: If you can't get your work done in a 24-hour day, work nights.

Parkinson's Law: Work expands to fill the time available for its completion.

Robinson's adjustment for the twenty-first century: When the available time contracts, the work continues to expand.

Stage 4) The Search For The Guilty

Well, the campaign was a disaster. The old biddies signed off in their hundreds. The heir lost his shirt, and his hair. The company was approaching its closing chapter. Where would everyone be landing, in lush green meadows or in the tar pit?

The lesson of the starlings

We should take another look at our starlings now, as they flock around Brighton Pier. Elegant and aesthetic as it seems, the flow of birds in a flock is not purely artistic, of course but due to the starlings' impulse to get to the centre, where they will be protected from predators.

I had noticed that one of the team, previously an amenable woman, allowed the trauma to get to her, and ended up constantly criticizing everyone else's work. What a negative influence she became; forever pointing out other's failings. Well, the truth is, she was trying to save her skin by pushing others out from the centre, where 'good, responsible, conscientious' people go, to the outside where 'incompetent, lazy' people go; building up enough alliances to either jump into the limelight if anything good happens or push someone else into the spotlight if it goes wrong. As the redundancy notices started to wheel and circle around them, the other staff began to follow her fashion. Ancient grudges, dating back decades to a different millennium, were dusted down and wielded against erstwhile friends.

Nothing could save the company. It would have folded anyway. But it might have been very different. They could have parted as friends, stayed in touch and supported each other in their future lives, as their ant side would have persuaded them. Instead they parted in bitterness and looked back in anger. I believe this rancour was not permanent, but that the ex-employees, having worked together for so long, were able to forgive and forget, and get back together eventually.

Stage 5) The Punishment Of The Innocent

The first head to roll was the idiot who got it right. She had mapped out the project's prospects on day one and spent the time quietly ticking off the blunders as they happened, sorrowfully shaking her head. Out with her.

The people who worked very hard went, because they were associated with everything, good or bad – and only bad counted right now.

People who did nothing went, obviously.

Stage 6) The Reward And Promotion Of The Incompetent

People who did nothing but looked as though they did lots were the survivors. Here is the secret of success – to disappear; like the Cheshire cat, the grin is there but little else. This is the art of 'reverse camouflage'. Camouflage means looking as if you aren't there when you are. Reverse camouflage is appearing to be there when you aren't.

Moriarty paid himself a big bonus, then negotiated generous severance pay with himself.

CHAPTER 7

PARKINSON'S LAW AND

THE PETER PRINCIPLE

The emblem of the ape-dominated world of work is the hierarchy. Orders pass from the top, where sit the plans and the experts, down through various levels of enforcers or managers to the lowly masses, whose only competence is to put things in boxes and make stacks. In the early days of the production line that's all they did. This didn't make them happy, but nobody cared; they were mere shelf-stackers, after all. Then it was discovered that if they were too unhappy, these lowly people could clump together and make trouble in the form of strikes, walk-outs or riots. Like slaves or prisoners they felt trapped; boxed up like one of their products.

It was not long before the notion of the meritocracy emerged; the idea that nobody was trapped at their level, but could rise through the system by hard work and intelligence. As they rose they would be paid more. This attractant produced the company 'sap', as people experienced at the lower levels add their knowledge to the management at a higher level, rising in some cases to be the top of the tree in the CEO's penthouse.

In the 1960s, Laurence Peter noticed an anomaly: the flow of talent seemed to be upwards only. Less competent people never flowed down the tree, back towards the stacking sheds. This led to his revelation, expressed in his landmark book, *The Peter Principle*:

IN ANY HIERARCHY PEOPLE WILL RISE TO THEIR LEVEL OF INCOMPETENCE

A law which, once uttered, is immediately intuitively true.

Promotion in these organizations is a carefully sequenced upward progression, from post room to team member to team leader to project manager to department head, and then onto a series of distant spirals to the upper echelons. Long-established institutions, such as the Civil Service or the Army, have charts you can view as you first arrive, with the salary scales at each level, responsibilities and benefits, and even the expected arrival time at each new rank.

Peter's theory states that the promotions will continue until the executive reaches a job which they can't do so well. Once having been promoted one rung too far, their upward trajectory falters. They won't be demoted – that would be too unkind – but they will go no further upwards. There they will stay for perpetuity: incompetent, unfulfilled, unhappy, disrespected, making life miserable for everyone in their department, but prevented by the system from escape.

Can we subject the Peter Principle to scientific analysis? Actually the scientists would say:

1) Your competence is a spectrum. Mostly you are adequate, sometimes you are rubbish, occasionally you have amazing insights. When you want promotion, which aspect of your work do you present to management? Exactly!

2) While you are going for promotion you will be striving extra hard, which you will cease doing once you've landed the job. You will be so exhausted from the effort of getting there, you won't have the energy for serious work for a month or two.

3) You will have to learn the ropes. This takes its toll.

4) People will be looking more searchingly towards you as the newbie. And they will pick holes, partly because they campaigned long and hard to get that job themselves and are exceedingly jealous.

More significantly for the Peter Principle, it is likely that you will be promoted into a new function that doesn't suit your particular skills. For instance:

1) The executive is promoted to a managerial position in a caring profession, where they need to be mild, soothing and full of the milk of human kindness. To land the job ahead of their rivals they had to demonstrate quite different skills – cunning, mendacity and belligerence. Some senior managers in the NHS fall into this category.

2) Perhaps the position requires understanding of children. The executive has worked such long hours to achieve the job that they never get to meet any. Not even their own. Executive producers in children's TV suffer from this. The most senior often have a photo of their children on their desk, not because they love them but because they can't remember what they look like.

Peter pointed out that this principle thrives best in a meritocracy, which rewards people for their ability. If promotion is earned through favouritism, nepotism, mutually shared hobbies or class, or if promotion is denied because of race, class, gender, sexual orientation, cut-of-the-jib or being called Dwayne, it becomes increasingly difficult to rise to a level of sufficient incompetence. For this reason alone we should be grateful for sexism, racism, Dwaynism and the public school system; because of them there are jobs being done by people who know what they're doing. If ever you have had an interesting conversation with a shop assistant or waitress, you are probably talking with a bundle of frustrations; someone who is desperate to make use of the MSc in archaeology they earned in Krakow University but can't because of the bigotry of the English. They won't be softened by the knowledge that they're doing excellent work where they are.

Not only must the hierarchy exist, but people have got to want to rise up it – it takes two to tango. What can be done with people who are happy where they are but not incompetent?

Teachers, for instance, have a habit of going into teaching because they like teaching. Conversations with excellent teachers show how hard it is to keep teaching if you are any good. Unless they cling to their classes with whitened knuckles, the currents will drag them away to somewhere they can be unhappy.

'Miss H, you are so good at your job, why don't you share your enthusiasm with trainee teachers, so they can be inspired and wish to emulate you?'

'Miss H, you are the only one who really understands what these kids need, why don't you head up the whole department? There's your cubicle. We'll pass the files through to you under the door.'

In conclusion, the Peter Principle may not be a rule without exceptions, but it brings into focus many aspects of the life of work in a hierarchy.

'The Ponzi illusion: Both these bosses are the same size (measure them), but the one further up the corridor of power seems more impressive'.

Parkinson's Law – Injelitance

Published eleven years before Peter, Northcote C Parkinson's book, *Parkinson's Law*, is a work of a dozen different sections, each one a little gem. More interesting than Parkinson's Law itself is the later section in which Parkinson introduces us to 'Injelitance', a mixture of incompetence and jealousy. While the Peter Principle claims that employees 'tend to' rise to their level of incompetence, as if the process is accidental – that incompetence wafts up through the system by a kind of osmosis – Parkinson's thesis is that once a manager has reached a level of incompetence, the only way he can stay there is by making sure there is nobody around to expose his hopelessness. He does this by promoting even more incompetent people into positions below him. They will in turn elect less competent people into jobs beneath them, and so the whole talent base will be corrupted. Competent people will be driven down and pumped out of the company, where they will set up their own, smaller, more competent businesses.

Normally the injelititis infection begins slowly, with the appointment of one injelitant manager into a junior position. From there his gradual promotion raises the level of incompetence higher and higher up the hierarchy until he can infect the central administration, and the disease is systemic, spreading to all divisions and regions. Sometimes the process is accelerated by the injelitant manager being propelled into a senior position from outside. As the result of a takeover, injelitant managers will be placed right at the top. This is either because the acquiring company wants to find somewhere to deposit a troublemaker at head office, or because he is the son of the new owner, who spends all day drifting around the house and needs to be given something to play with.

However, the infection takes hold, the process is inexorable, and nearly all companies drift down towards a protracted death, as competence leaks out at the bottom, forced down by managerial peristalsis.

Cases of recovery have been known, owing to a rare antidote. Some people are so clever at hiding their intelligence that they can pass unnoticed up through the ranks (helped, of course, by the systemic stupidity of the organization). They can allay suspicions by periodically comparing ski holidays or harping on about violations in dress code at the golf club. Once they have achieved the top spot, they can throw off the mask and appear 'like the demon king among a crowd of pantomime fairies'. Their incompetence suddenly exposed, the whole management structure will hastily stuff wads of cash in their pockets and run for it. From then on recovery is possible.

Parkinson's Law and the Peter Principle both apply well to a world of hierarchy. Our world view is dominated by the career structure, which rewards hard work with greater responsibilities and the opportunity to see even less of your family. And when you die and go to Heaven you will find another hierarchy. The divinities have been ranked since the Middle Ages into Angels, Archangels, Principalities, Powers, Virtues, Dominations, Thrones, Cherubs, Seraphs, the Left Hand of God, the Right Hand of God, and the CEO himself. Presumably these positions are also filled by incompetents, to make the blessed souls feel thoroughly at home for all eternity.

SECTION 3

ANATOMY OF CHAOS

THE ORIGIN OF SPECIOUS

When I was a student I took a holiday job as a labourer on a motorway being built near Bristol. I believed that since motorways were such sleek, smooth things, the building of them would be a sleek, smooth thing too. Well, perhaps I might be forgiven; I was very young and had little experience of the world of work. Every single job had to be done at least twice. Firstly in the wrong place, then in a different wrong place, then in the right place at the wrong time . . . I became very attached to a row of kerbstones I carefully positioned, following the line of the slip road. They were all knocked down by the lorry dumping earth behind for the grass verge. They were re-erected, then levered out again to be placed somewhere else, then repositioned further over after somebody revised the plans again. If you drive off the sleek, smooth M5 at junction 21, wave to my little section of kerb. You may imagine that it was built efficiently. Not a bit of it; it was much more fun than that.

The various apes who drew up the plans wrong, or read them wrong or scheduled them wrong, were helped by the myriad 'ants' like me, who plodded around in the mud trying to do our best, following the guys nearby without having the first idea what it was all about. This is, was and will be the way life is lived all over the planet.

This section is devoted to chaos. Chaos is created by social fractals, when one idea collides with another. We find the results littering the tabloids and gossip columns under 'you couldn't make it up' headlines: 'Health and Safety Gone Mad', 'Bonkers Bankers', 'Crazy Council Strikes Again' . . .

Evolution is also driven by chaos. Evolution works through random mutations, some of them very random and very mutant. During the Cambrian era, about 500 million years ago, an enormous range of new species evolved, many of which were completely useless, and died out rapidly, before the *Cambrian Times* could announce 'Wonky Worm with Twin Heads Starves Between Two Meals'. The reason we know they existed is that their bodies sank to the bottom of the sea and were covered in a particularly fine mud in a place later to be called Burgess Hill. Five hundred million years later, their fossils were accidentally stumbled upon by a descendant of another, more successful branch of evolution by the name of Charles Walcott, and then we began to discover the enormous range of oddities that evolution created, then discarded. That chaotic burst of evolutionary zeal was called the 'Cambrian Explosion', and the fossil site, the Burgess Shales, became a sensation. The thousands of impossible creatures found there were given names like *Anomalocaris* ('unusual shrimp') or *Hallucinogena* ('hallucination').

Today, around us lies the debris of another explosion of unusual and hallucinatory schemes and dreams, products of our cultural evolution. Our rubbish dumps are the Burgess Shales of tomorrow, stuffed with extinct objects with weird names like 'typewriter', 'carbon paper', 'slide rule', 'LP', 'fax'.

Along with these solid remnants we are surrounded by the less solid rituals of today, which our children will laugh at tomorrow. Let's pre-empt that by laughing at them now.

Our social fractals spin off into whirls of absurdity as strange and unlikely as the Burgess Shales. In particular, two interesting whirls: positive and negative feedback loops.

Chapter 8: The Law of Unintended Consequences: Negative feedback loops happen when self-contradictory behaviours cancel each other out – when the left hand doesn't know what the right hand is doing. For instance, on one day in 2007 the British government published plans for cutting greenhouse emissions from aircraft, while on the same day approving plans for the expansion of Heathrow Airport.

Chapter 9: Run-Away: When a simple idea reinforces itself through a positive feedback loop, the result is an escalation of craziness, as when a parent smacks his child to stop it crying.

Chapter 10: Creativity: Sometimes our mutations turn out to be improvements. This is the basis of all human progress, and why we must tolerate chaos. We must trip over our feet a thousand times before we can take one firm step forward. For creativity's sake we must endure everything, however insane. 'Creativity' is the name we give to our more successful blunders.

CHAPTER 8

THE LAW OF UNINTENDED
CONSEQUENCES

In order to finish off a late-running project, a company in Madrid took the rational decision to take on extra staff. Of course they needed to be trained to do the job. The training process would be three weeks, and people were removed from the production line to help do the training. Because of this, the production line began to go more slowly. They were going to need more staff to make up lost time, so more were hired. These had to be trained, so more people had to be taken from the production line. So more had to be hired to fill the gap in the project. But now they had to be trained, so . . . Mysteriously, the more staff they took on, the slower the production line went. Here is the negative feedback loop of the Law of Unintended Consequences, reminding us that however well rounded our plan, sooner or later it'll go pear-shaped.

Tell me and I'll forget; show me and I may remember; involve me and I'll understand.

The teaching of science has wevolved into the teaching of how to pass science exams. A bad science teacher confessed in *New Scientist* magazine: he had been a good teacher once. His students learned to investigate, work out theories, test them, and describe their findings in their own words. They understood their subject perfectly. The unintended consequence was that they failed their exams. The reason was that those who marked their papers had a checklist of words. They gave a mark when the key words appeared. If the student described the topic perfectly, yet failed to use the key words in the correct place, he was marked down. Nowadays, says the bad teacher, he gets better results; he spends his lessons dictating key words carefully to the students. They pass their exams. They don't have any idea what it's all about, but they have the certificate, and that's all that matters. They are happy, their parents are happy, the school is happy. Only the teacher is miserable.

Kids like danger, and growing obese is very dangerous

Australian health agencies launched a health campaign in 2004, encouraging children to cycle everywhere, to help deal with a growing problem of obesity. The campaign worked, and the streets were filled with cyclists. To keep them safe as well as fit, the Australian government made cycle helmets mandatory. The unintended consequence was that all the kids decided the helmets were too uncool, and went back to being couch potatoes.

Sometimes your guardian angel is your Big Brother

Workers for two of Camberwell Council's transport departments made interesting times for resident Ruth Ducker. She had always legally parked her Volkswagen Golf around the corner from her house in an unrestricted street, so it came as a shock when she discovered it had disappeared from its spot, and in its place were some double yellow lines. After three weeks searching, the council admitted they had taken the car, and told Ducker she owed more than £800 in fines. In fact, the car had been carefully lifted out of the way so council contractors could paint the double yellow lines, then replaced. The same day a different set of parking enforcers spotted the now illegally parked car, and had it towed away. It's ironic to note that nobody did anything wrong.

The road to hell is paved with good intentions and cycle lanes

On the one hand the Transport Department want cars to circulate untrammelled round our towns, on the other hand the council's cycling officers want more cycle lanes for the safety of cyclists. They lobby hard with the powerful officers in the Transport Department, and the result is a lopsided attempt to keep everyone happy: the 5-metre cycle lanes. They appear; seconds later they disappear, casting you back into the traffic. One couldn't devise a better way to annihilate cyclists.

Money is much more exciting than anything it buys

A major British bank had a brace of policies in their branches which rubbed wonderfully against each other. The branches were given bonuses for the amount of cash notes they didn't hold overnight (the argument being that the more hard cash they didn't hold, the more money was available for the bank to earn interest on over the night hours). At the same time they were given bonuses for the hours they could keep the cashpoint operating through the night. So the bank was hoping to be as generous as possible and as greedy as possible at the same time. The problem was that the cashpoints needed notes, so in earning the bonus from one scheme the bank would lose out from the other.

The Left-hand/Right-hand Law ensured they were damned if they did and damned if they didn't. One branch manager's elegant solution was to put small amounts of money in the cashpoint, but crumple the top note. The machine immediately broke, which meant that it was defaulted to 'open all night' status, while not needing any notes to operate, thus earning both bonuses.

When it comes to being over-fastidious, you can't be too careful

A company's stores manager who went on an excellence course was taught to keep a safety margin; to order new stock before he was out of the old stock, so he could keep the shelves full. Some weeks later someone asked him for a component, only to be told it was out of stock. 'But I can see one there on the shelf,' said the applicant. 'Can't touch that,' said the stores manager. 'That's the safety margin.'

Always expect the unexpected

I'm sorry to have left out so many fine examples of the chaos humans inflict on each other. Young people look forward to the future, when life will be perfect. Old people look back to the good old days when life used to be perfect. Both have to live in the present, which is chaos. And it doesn't get better when we study run-away...

CHAPTER 9

RUN-AWAY

I was a skinny young thing when I took a part-time job for a year at a charming local school, where everyone seemed to compete to be more generous than each other in the way of break-time snacks, bringing ever larger bags of sticky buns each day. By the end of the year I was a stout fellow.

When anyone left the staff or had a celebration a greeting card was purchased and everyone signed it. Over the year there was a measurable increase in the size of the card. A card shop had opened in the town, offering a suddenly new, much wider range of cards. From the traditional postcard size, our cards grew over the year to be as big as tea trays.

Why do cards keep getting bigger? Why do cakes and biscuits get bigger? Why do waistlines get bigger and bigger and bigger? Run-away is the answer.

'Run-away' was originally used by evolutionists to describe seemingly absurd occurrences in nature, like the peacock's tail. How could something as peculiar as that tail exist? It is so fabulous when up but so cumbersome when down, only a maniac could have devised it. The process was actually long, slow and deliberate. It began three million years ago, when peacocks' girlfriends started to select males with better plumage, and so peacocks with larger collections of fine feathers found themselves more in demand than their more modest chums. These tail designs got passed down to the next generation, where there were more peacocks with big feather displays. Occasional mutations produced offspring with even more attractive tail feathers. The more the tail displays were sought after, the bigger they became, therefore the more they were sought after . . . Natural selection made the tails evolve out of all proportion. Run-away ran off the cliff the moment the tail became so heavy that it threatened the survival of the species. When he isn't courting, the peacock drags his display around like a tablecloth stuck to his bum. The only reason he survives is the rarity of predators in his habitat.

Run-away runs away quickly among humans. We don't have to wait for evolution to create our absurdities as the peacock does; we can do it in a single year, with wevolution. The expanding size of the leaving card is explained: each new 'generation' of leaving card is selected for the size of its display. It rapidly evolves bigger. The break-time sticky buns in the patisserie are also selected for the size of their display: icing, raisins, glacé cherries and how many hundreds and thousands. They grow bigger and bigger. As a result, those who eat them grow bigger and bigger and bigger, to apparently no evolutionary disadvantage, due to the lack of predators.

As we will see later, glorious displays are something humans enjoy evolving on pretty much a daily basis.

Run-away finances

Financial institutions have wevolved over half a millennium, since Columbus discovered America and trade began to expand around the world. Like the peacock's tail, banking has grown complex and exotic over the years. Run-away has moved from cash trading to futures, hedge funds or derivative swaps. The language suggests some kind of mating display; they offer customers more 'interest' to attract them. Rival investment companies gather round with ever more lustrous displays of financial instruments and interest rates, riffling them in a shimmering flourish, hoping you will be interested enough to let them bond with your money. Greater offers of interest here inspire even greater offers there, until the interest rate has grown so huge that, like a peacock's tail, it is unsupportable; a predator bank swallows up the cumbersome company and the investors' deposits become extinct. Between 2008 and 2011, 414 American banks became extinct because they had promised too much to their clients and couldn't deliver.[1] At the height of the crisis my bank phoned me up to ask if I wanted to switch to a higher-yield account. This was the very thing that had brought the banks to their knees in the first place. Naturally I said yes. Our own greed feeds the banks' greed, so we all end up in it together.

As the waves of boom and bust follow one after the other, they show self-similarity, as any social fractal should; each bust is preceded by wild optimism, suicidal financial chicanery and blithe assurances, and followed by bewilderment on the faces of our wise leaders as the wave breaks. An early example of these 'bubbles' was the great tulip mania of 1636–37, which gripped the investors of Holland in frantic price run-away when they fixed their beady eyes on the humble spring flower. Anything can become valuable if everyone decides it is so. And so it came to pass for the most unlikely investment. At one point a single tulip bulb was valued at the price of a small farm. At another point shortly after, when the bubble burst, it was valued as so much compost. When bubbles burst, the pendulum swings the other way and we experience what the city calls 'correction' and we call 'poverty'. Financial bubbles burst every 15-20 years. You can bank on it.

Automated run-away is now available online with, for instance, booksellers like Amazon. Sometimes books are priced at millions of dollars ($23,698,655.93 is the leader as I write, for a book about fly genetics[2]). How come? To sell your book as a retailer on Amazon you should either be the best retailer or the cheapest. To be the cheapest is easy: you install software that checks all the book prices and make sure yours is a penny cheaper, to get you to the top of the list. But if you are a seller with a better reputation you could charge 10 per cent more than that. Unfortunately, you don't have the book. So you advertise it anyway, at 10 per cent more than the cheapest, wait for someone to order it from you, then buy it off the cheaper guy and sell it on. The problem is that the cheaper guy's software will automatically re-price their book at a penny cheaper than yours. Your software will re-price yours at 10 per cent up from theirs, and the price spirals to giddy heights within days.

You can sometimes find run-away in company entrance halls. It's good to impress clients as they arrive at your offices, but some companies do lobby bling – glorious façade, sumptuous entrance atrium, fountain, goldfish – sometimes even *gold* fish – and a receptionist fresh from Cheltenham Ladies' College. The advice from investment analysts is that if you see this in a company you have invested in, withdraw immediately. They are suffering from image run-away, and are probably going to run away with your money.

Run-away technology

If something can runaway, it will. Wevolution is like a dog on a leash, waiting for the chance to run amok.

Security
How safe is your personal online information? How many permissions do you need before you can get to it? Security run-away can sometimes keep your identity even from you, to be on the safe side. You know your information is properly secure when you can't access it yourself.

It's a romp for IT experts to insert security firewalls between every application and database, each one operated with a different password, user name and security number. This is the security industry's answer to the peacock. We all want security, and now we have oceans of it. Big companies that handle sensitive information have reasons to be careful, but run-away means that small companies or even households can find themselves with as many passwords as the Pentagon. If you follow the advice of security experts you never duplicate passwords, many will advise against choosing a password that is memorable and nearly all recommend you change them regularly. They have become impossible to remember, so the only solution is to write them down on a Post-it note stuck to the screen. Which rather ruins the point.

Online User Guide

Your online software manual says everything is so easy. 'Just click and go' invites the message on the screen, followed by a simple set of three options, each with three further options, each with menus of a dozen items, each with a sub-menu of five or six, so that in no time 300 possibilities have been ranged out before you; you, with your short-term memory capacity of seven. The on-screen guide has wevolved into a magic pasta pot – it just keeps growing. Nobody knows how big it is because you can only see one page at a time. In the days of the printed book, not so long ago, one could flick through the pages, stick a finger in one spot while looking for another, and in the end one knew where one was. A book has a beginning and an end, and an editor desperate to keep the bulk down, but the on-screen guide has menus and sub-menus and links to links, loose 'pages' that float unbound around the ether. One software site I visited had over 30,000 pages and links. Home pages are no longer homely, either. Many have evolved into a peacock's tail of buttons, flashes, offers, logos and neon glitz. The *New Scientist*'s home page currently has over 120 separate buttons, the BBC's 212.

Answering machines

Answering machines are designed to be personalized by each company that uses them, so they are packaged with the full Pandora's box of absurdities ready to be sprung on us. So, for instance, you want to complain about a gizmo you bought in Essex. You phone them up.

'Thank you for calling . . .' says the voice of a bright sparkling American lady.

'. . . Canvey Hardware Solutions . . .' continues a nervous estuary bloke.

'You are being held in a queue . . .' the American lady adds.

You may not have noticed it, but inadvertently you have clenched your fist. Not because you are being 'held in a queue', not because your call 'might be recorded for training purposes', not because the announcer seems to change nationality and gender a little too often, but in anticipation of the music. The people who set this up have a taste in music shared by about 30 per cent of us. That means 70 per cent hate what follows: Vivaldi's Four Seasons, someone fumbling on a stylophone or being dragged to a Dire Straits concert, if not an extended advert for the company whose product has just broken, which is why you are about to spend half your morning in a queue.

'Your call is valuable to us'. At this moment you may spot a paradox: the voice at the other end is a machine. Can a machine really 'value' a call? And who are 'us'? Are there lots of them out there? Is this a breeding colony of answer machines?

The automated answering service exists in its own evolutionary niche, unreachable by humans. It can't be touched by the incoming caller because, however much you shout at it, it will only say, 'I'm sorry, I didn't quite catch that. Please tell me the number of the extension you wish to contact'. Neither can it be influenced by the person you are calling since they never hear it, so they are unaware of the problem. In this uniquely isolated, dark habitat, the answering machine can mutate. Extra voices with alien accents can add their pennyworth: 'Please choose from the following thirteen options . . .'. The more they try to humanize themselves ('Right, to get the ball jolly well rolling, let's press the star button, shall we?'), the more unnerving they become. Their professionally charming voice will talk right through you, will tell you with the happiest of smiles that your extension doesn't exist, please try again later, you are now exiting the system, thank you for calling, click.

Software

A software application that calls itself easy is as trustworthy as a man who calls himself honest. When a perfect, simple and effective software application first comes on the market it is always a pleasure to play with. But a year later, because the manufacturer wants us to buy an improved upgrade, extras are added. By the fifth upgrade run-away has ensured that any new buyer is flummoxed. It cannot be understood unless you go on a three-week course. (Can I mention Microsoft Word at this point? I am typing this using Word, so I ought to be discreet. 130 items in the drop-down menus at the top. One of these, Preferences, itself has 130 entries. Word was at its best on version 3. By now any attempt to do more than type a letter is doomed to failure. A peacock among programmes.)

While the peacock-ware is being developed, bugs will slip in owing to the different designers all using different codes somewhere down in the software's engine room. These should be ironed out before putting the product on the market, but that takes time. A much smarter idea, now widely adopted, is to launch the product with full marketing razzmatazz, plus a goodly bunch of errors, then allow the public to find out the problems themselves and phone up your helpline, then you can send out updates which deal with all the issues. The benefits are obvious. You don't have to pay for professional bug hunters, the calls are all on the customers' phone bill, and after a little while you can switch to premium phone lines and actually earn off them, so the more bugs there are, the jollier it all becomes. Everyone is a stakeholder in the software, and the final programme is an amalgam of everyone's efforts. This is close to the system used by ants, with everyone chipping in to help build the nest. Only ants don't get phone bills.

Run-away health and safety

This is the age of expanding health and safety. We should relish it, because safety is an expensive preoccupation, and may not provide us with amusements for too many more years. We all want to be healthy and safe, and there are certainly more dangerous machines around us now than ever before, but we can't help running away with ourselves sometimes. H&S officers can push the envelope a little more each year, because any suggestion that they are being a little too fretful with their precautions comes over as a willingness to endanger life. The positive feedback loop comes from everyone's enthusiasm to show themselves more zealous than everyone else.

Pupils get taught while teachers become tauter

The full flowering of Health and Safety occurred at Hatfield Primary School in Merton in 2006. The council had hired a lollipop lady to help the children cross the busy road outside the school at the end of the day. On one particularly windy day, the council-appointed manager of the lollipop ladies decided they ought to stay at home because their lollipops could be a hazard if the wind caught them and blew them into people's faces. Faced with an absence of lollipop the Hatfield School teachers were in a quandary: they couldn't help the children across the road because they hadn't been certificated to do it. So in the very best interests of Health and Safety, the children had to cross the high street unassisted.

Safety doesn't happen by accident

In April 2007 at the Health and Safety Executive's Headquarters, it was decided that only qualified staff should move chairs for fear of injury. Signs appeared warning staff: 'Do not lift tables or chairs without giving 48 hours' notice to HSE management'. Back strain is a major cause of injury in the workplace, possibly because staff don't get enough exercise. Not being able to shift chairs can only add to the problem.

Tear-away

*The only way to find out how far is too far is to go
a little bit further*

The despair of Health and Safety officers is that we don't appreciate them; we want to take risks. When schools started throwing out playground climbing frames, for Health and Safety reasons, they found the kids climbing over the walls, trees and dustbins instead. Nothing can stop the urge to climb. If we aren't allowed our run-away, we will tear away.

Some people have to take things to extremes. We call them groundbreakers or idiots, depending on whether or not they survive. If we are going to keep going forward, then we have to tolerate those who go that little bit too far. So people will climb a mountain 'because it's there', or throw themselves at the Moon because there's nowhere left to explore on Earth. Nearly all scientists have had a go at blowing themselves up; it's practically obligatory. Robert Hooke swallowed all sorts of quack remedies and noted down in lurid detail what damage they did to him. Newton pushed a spoon behind his eye to see what he would find. Dr Barry Marshall deliberately infected himself with the bacterium *Helicobacter pylori* to prove that it caused stomach ulcers. (He earned the Nobel Prize for his recklessness.)

Tear-away was best seen at Chernobyl in 1986 – the 'I-wonder-what-happens-if-I-turn-this-knob-just-a-little-bit-further' syndrome was applied to a most unforgiving material: uranium. Nuclear power is the most complicated, the most dangerous and the most expensive way to boil a kettle that mankind has yet devised, so you have to be very careful with the handling of it – how careful is difficult to ascertain without edging ever further into the danger zone. Tests were proposed to probe the safety systems of the Chernobyl reactor number four on 25 April 1986. No doubt the nuclear scientists who devised the tests were fastidious in their preparations, but in a moment of sublime recklessness they then handed the operation of the experiment over to the night shift, who had very little idea about nuclear power, having been employed in coal-fired power stations all their lives, and through a string of errors they blew up the reactor and made uninhabitable an area of Ukraine the size of London. So that taught us something.

The problem is, nobody can know the limitations of safety systems except in the past tense. It is impossible to predict the future until after it has happened, so there have to be errors in order to give us the knowledge necessary to plan for them. There is a certain dissatisfaction with safety measures which are successful. The SARS outbreak of 2003 was contained successfully by enforcing strict quarantine and border controls – or was it? It is widely believed today that there was no risk. The 'proof' is that there wasn't a pandemic. But how will we ever know if there might have been? Perhaps the precautions worked. The Millennium Bug was going to devastate our computer systems. In the end it didn't, but can we state categorically (as we do) that there was never any danger? Might it not be that as a result of the widespread alert some vulnerable systems were usefully adjusted? Nobody really believes that. They need positive proof, which is something you only get if you let it all happen. For that reason we should confidently expect, in the next decade, another nuclear accident every bit as educational as Chernobyl.

What if Health and Safety retroactively cancelled all past human endeavour which involved risks? The space programme is the first to go. Wars will be stopped, of course – even with conkers. No science, because of the risks of poisoning, explosions and bruised fingers; taking things further back, no sailing, no playing with fire, no bashing flints together (think of the sharp edges). So far we've cancelled civilization as far back as the early Stone Age, but it doesn't stop there. Let's not have you swinging from trees, young *Homo erectus*. And what's that in your mouth? Are you sure it's edible? Have you cooked it? Did you get permission to light that fire?

Sadly for our health and safety, the risk-taking gene must stay. It is the one that moved us forward from the amoeba. I suspect that if H&S had been around 4 billion years ago, the flagellum would have been banned.

So we need to make mistakes; they are the stuff of creativity.

CHAPTER 10

CREATIVITY

Creativity is a special kind of error. 'Error' comes from the Latin 'to wander'. How can you think of a new idea except by wandering away from the old one and getting a little lost? To be creative you have to kick your mind off the normal path and into the undergrowth. If your thoughts know where they're going, you aren't being creative yet. All these items demonstrate extreme ape; a dogged determination to break the rules.

The Xerox Corporation made a giddy fortune from photocopiers and colour printers, so much were they part of the fabric of life that we still talk of 'xeroxing' something when we want a photocopy of it. In 1970, they opened a research facility to develop new product lines in Palo Alto, California. The head office in New York became convinced that it was a waste of money. The scruffs who peopled it spent their time hanging out, chatting, playing games and doodling. Not only that, when they took a look, Xerox found that nothing could be done with the inventions. A 'mouse'? A 'graphical user interface'? Xerox decided there was no commercial potential in them, nor in the so-called 'personal computer' gadget. A young visiting loafer called Steve Jobs thought these ideas had potential and started making his own products from them. Xerox were so bound up in bureaucracy that they didn't even manage to sue Jobs when he used their technology in his Apple Mac, released in 1984. Moral: to err is human; to profit from it, divine.

Creativity is at the heart of progress. But you wouldn't think so from listening to managers. Every boss hopes that the creative bit will be done some other time or on someone else's budget, maybe at home* in the evenings, so that during office hours they can get on with doing their proper job smoothly and efficiently, on time and to cost. Creatives are a challenge. They are for ever making suggestions which threaten the routine.

Everyday creativity

The printer that sits on your desk used to be the size of the whole desk. It would have cost thousands and worked slowly. Now it costs a few pounds and works brilliantly.** It handles paper sheets one tenth of a millimetre thick, teasing out each one in turn and steering it with pinpoint accuracy through

---

* When it comes to the financial benefits of creative thought, however, most bosses will ensure that the reward is reaped by the company, not by the individual.

** Meanwhile other creatives have been refining the ink ransom – you pay nothing for the machine until you want to use it, then you find each ink cartridge is half the cost of a new printer.

a box the size of a biscuit tin. Its evolution from dinosaur to dinky delight was not achieved by faithful copying from one generation to the next. Nor was it done by a few brilliant inspirations, but emerged by a process of hunch followed by tweak, performed by hordes of unknown creatives who twiddled. The results of their wandering ideas are that the printer gets better as the years go by. If you look inside you can see hundreds of tiny springs, rollers and cogs, each one meticulously refined to give exactly the right amount of pressure in the right direction at the right time, to print your document in full colour, with pinpoint accuracy.

Catering creativity

The neatest demonstration of the value of errors is in cooking, where the resistance to change is normally highest, because we are so very fussy about our food. Changes to our diet tend to be very slow in coming, unless something kicks us into creativity.

The Bakewell tart should have been a disaster. At the White Horse Inn in Bakewell, Derbyshire, sometime in the 1860s, Mrs Greaves, the landlady, asked an inexperienced kitchen assistant to make a strawberry tart. The assistant, however, left out the eggs and almond paste mixture in error. To try and rescue it, the eggs and sugar were used to make a filling for the plain pastry case, spread on top of strawberry jam. Voilà! Not a disaster but a sensation.

Chicken tikka masala was invented in Glasgow a century later, when a customer who found the traditional tikka too dry asked for some gravy. The chef could not bring himself to stoop so low as gravy, but improvized a sauce from tomato soup, yogurt and spices, and the Tikka Masala was borne out of adversity.

The potato crisp arose from a fit of pique. George Crum was a chef in a small diner in Saratoga, New York. He wasn't very good at cooking, but he was awful at handling complaints, so nobody grumbled unless they were angling for a fight. In 1853, one of the customers sent his chips back twice because they were underdone and too thick. Crum exacted revenge by slicing the chips microscopically thin and frying them so much they shattered when the customer tried to spear them with his fork. Instead of storming out in a rage, as Crum hoped, the customer ordered more.

Problems with creatives

Creatives get it wrong

Getting it wrong is an essential part of getting it right. How can you find out what works unless you dig through what doesn't work? To find one diamond you have to dig out two tons of rock. The same goes for good ideas.

Sturgeon's Law states: 90 per cent of everything is crud.

We can all nod our heads to this, those of us who work in industry, the construction business, TV or building ants' nests. Creatives would say so too, but they would suggest the figure is a little modest. Only 90 per cent?

Soichiro Honda said: 'Success is 99 per cent failure.'

And when Thomas Edison said: 'I didn't fail, I found ten thousand ways that didn't work,' he must have been hinting at – good grief – 99.99 per cent crud; pretty close to the observed value.

The history of medicine is thick with lucky errors. People in pain will try anything. In days gone by, if your headache was agonizing enough you could get some relief from sinking your teeth into a tree and chewing. If you were lucky enough to chew a willow tree you would feel, unexpectedly, a whole lot better. The bark contains salicylic acid, eventually extracted to make aspirin.

It was a failed attempt to make another drug, quinine, which produced a lucky accident. In 1856, William Perkin was cleaning out his equipment after another failed experiment when he noticed a pale purple colour spread across the sink. He had accidentally created the first artificial dye, a brand-new colour for which he coined a brand-new word: 'mauve'.

Creatives need errors

Alexander Fleming kept a famously untidy laboratory, with used equipment lying uncleaned on every surface. It was because of his lack of hygiene that one of his Petri dishes became contaminated with a wayward fungus floating in through the window. But it was because Fleming had a prepared mind that he could see the possibilities in the error. Instead of throwing the Petri dish away, he examined the effect the fungus had on the culture in the dish. It had wiped the bacteria out. Fleming realized that the fungus, penicillin, had a great future.

Creatives fiddle with things that should be thrown away

George de Mestral should have sent his trousers to the cleaners in 1941 when he found the turn-up covered in burrs after a country walk. Instead, he fiddled with them, worked out how they clung so effectively, and invented Velcro.

About 6,000 years earlier somebody was found checking through the ash at the bottom of a fire, instead of clearing it away. What they found there, molten lumps of metal that had dripped out of an ore-bearing rock in the fire, moved us out of the Stone Age into the Bronze Age.

Creatives waste time

Charles Goodyear spent a lifetime trying to solve the problem of perishing rubber. Natural rubber was always known to be wonderful – light, bendy, cheap, waterproof, easy to mould into interesting shapes – but it tends to go soft in the heat and brittle in the cold, and it perishes. Goodyear blundered around for decades, mixing different materials with it in the hope that one of them would make it more stable. Every time he thought he'd cracked it, he brought out a series of products made from the new wonder material, won prizes, sold tons of goods, then had to deal with the sticky messes that were returned a month later by angry customers. He got heavily into debt. At one point he thought sulphur would do the trick. When that experiment failed as well, he threw the kit on the fire in a fury. Next morning, clearing out the fire, he found that the rubber had gone hard in the heat. It was a discovery that changed the world. His next blunder was to call it 'metallic gum elastic' – quickly rectified to 'vulcanized rubber'.

Creatives are lazy

Richard Feynman, winner of the Nobel Prize for physics in 1965 for his work on electron spin, was another creative. Some of the inspiration for his ideas came from loafing about in a café in Cornell University. Feynman had been depressed for a long time after the death of his wife, and loafing around was just about all he could do. Some students in the café were loafing in their own way, spinning plates across the room. Idly, Feynman worked out the relationship between the spin of the plate and the amount of its wobble. He realized that this relationship could be applied to electrons as well as canteen plates, and it was the inspiration that led to his Nobel prize-winning work. If Feynman hadn't taken 'down time', he wouldn't have had the inspiration.

Johannes Gutenberg was hanging out at a wine festival, taking a break from thinking about how to build a printing machine, when he saw a wine press in operation. He realized that this was how to do it. The rest is history, and history books, of course.

Would we have had 'eureka!' if Archimedes hadn't decided to stuff it all and have a bath?

Creatives are undisciplined

They don't sleep proper hours, they eat badly, sit wrong, don't wash, leave their rooms in a mess. They can sit, suppurating, for days in the corner apparently doing nothing. Or so the landlady says. But this is because they are focusing on the one, the only problem that is important to them. Lots of useless slogging at a problem is a necessary prelude to the single 'ping' that solves it.

General Eisenhower, who masterminded the invasion of Nazi-held Europe in 1944, understood how important it was for even soldiers to avoid too much discipline. He said, 'In preparing for battle I have always found that plans are useless, but planning is indispensable.' Eisenhower gave the platoons freedom to change their plan according to the circumstances they found on the ground, circumstances no general miles behind the front line could possibly predict. Battles tend to be rather unpredictable affairs. Nobody can be sure where the next grenade will come from. It would be crazy to go into a battle with no idea about the terrain and enemy strength, but to have too rigid a plan would be equally reckless. The set-piece battles of the First World War saw well-disciplined lines of soldiers march punctually towards the enemy bullets. The first assault of the Battle of the Somme, on 1 July 1916, was a logistical masterpiece: 20,000 soldiers died.

Creatives are essential

As I sit here I find that within arm's reach I have over 400 inventions, ranging from the desk-top printer sitting beside my computer to the paper in my notebook, invented by Ts'ai Lun in 105 AD. The ballpoint pen, mobile phone (itself housing more than 1,500 patents), postage stamp, the shirt I'm wearing . . . even the apple core mouldering to my side was invented by humans – or at least bred over thousands of years from a small, bitter Asian berry. It's a wonderful, rich, diverse world we have made, and we would never have made it if we ever thought for one moment that we had got it right.

The reason why we have such a plethora of inventions, while ants are still stuck in the mud, is that they evolve, while we wevolve. To get from Gutenberg to ink-jet took us 550 years. If it had had to evolve genetically, it would have been more like 55 million years.

For every good idea that survives today there were hundreds that were brilliant in their day, but have faded away. Before modern printing systems there was the woodblock, intaglio, etching, lithography, movable type, rotary press, hot-metal typesetting, the golf ball . . .*

Somewhere among the insane practices that surround you at work may be the germ of an idea which will evolve to bestride the Earth for centuries.

---

* The IBM Selectric typewriter of 1961 substituted the 'basket' of individual type-bars that swung up to strike the ribbon and page in a traditional typewriter with a single 'golf ball' covered in letters rotated and pivoted to the correct position before striking. Brilliant! It lasted 20 years before being added to our landfills.

SECTION 4

APE OR ANT

CHAPTER 11

CAN YOU TELL THEM APART?

Apes and ants seem very different: apes battle for what they believe to be right, they campaign, they feel passion, belief, pain and joy. Ants follow their neighbour. They harmonize. They just don't do controversy. They say, 'Pain is inevitable, suffering is optional'. (Haruki Murakami)

The inner ant says, 'Together we can do good things.' The ape says, 'We can do things a lot better if we work out a management tree first.'

The ant says, 'We are equal, we multi-skill.' The ape says, 'We specialize, we have different ranks.'

We have met the ant and the ape when we filled out the job application. The ant told us to write: 'I enjoy working as part of a team . . .' And the ape added: '. . . but am capable of using my initiative to delegate'.

Here are some typical situations that create ripples in the normal office life. (For many offices 'normal' is a somewhat rare event.) Relish the struggle between ape and ant.

157

The stapler

Lives have been lost over staplers. You put it in a safe
place; everyone knows it's yours, you turn your back for
one second and whammo, it has disappeared. What do you
do? Stand in the middle of the office and shout, 'Which of
you bastards nicked my stapler?', like your home has been
burgled and your children sold into slavery? That would be
the ape speaking. Or do you take the ant's approach, tell
yourself that everything flows, including staplers, and pop off
to nick someone else's?

Troublesome colleagues – one
The urge to harmonize and work together like an ant is
strong, but the urge to massacre the guy on the desk next
to you if he says 'No problem' one more time comes a
close second and is catching up fast. The problem for our
inner ape lies in the size of the office. Real apes live in
the open air, so they can put plenty of distance between
them and the 'no problem' problem. When you're sitting
shoulder to shoulder all the time you get tense. Surprisingly
small things can get to you: the clicking of his biro top,
the tossing of his hair, his particular style of sigh. With
no escape except throwing yourself out of the window
(and many offices have installed ingenious catches on the
windows nowadays to stop you doing that), it can raise your
anger level and lower your threshold for showing it. It starts
with a glance, goes on to tuts and grunts and can end with
physical violence if your neighbour doesn't get the hint and
change his behaviour.

Troublesome colleagues – two

The guy on the desk next door who has anger-management difficulties. It starts with a glance, goes on to tuts and grunts and can end in physical violence if your neighbour can't control himself. No problem! Simply mention it to HR. They'll sort it out, no problem.

Anecdote rage

When everyone returns from holiday is there a competition to see who can tell the most outrageous story? Do they book into hotels without roofs, whose swimming pools double as cesspits, simply so they can regale you with long screamers about dysentery? Do you then have to reach for an even worse experience, which you had fifteen years ago, simply to top them? Do they have 400 emails waiting for them? Does that signify their importance? Must you now sign up to a spam service so you can outdo them?

The cubicle

The cubicle was invented firstly to satisfy our desire to defend our personal space, and secondly to pander to management's desire for fewer murders. Now a third benefit has been found: in a cube farm, operatives can be stacked to the roof, space maximized, rent optimized, problems disguised, territorial claims marginalized.

Special conditions pertain in finance houses. You might think that city trading firms, who make millions in a day, could afford to give their traders their own offices, with a secretary and a golf course. But no, traders' desks are incredibly close together, the distance set precisely to maximize aggressiveness. Research has shown that aggressive traders do better deals, so management deliberately sets out to create a powder-keg atmosphere in the trading rooms by piling the testosterone balls into the broom cupboard. In spite of this, for many years economists spoke of the 'rational market' as a natural dampener of trading excesses.

The absence of cubicle

When the new office manager decides to go hot-desking and open-planning, with executives wandering the floor like ants looking for a patch of sun, do you welcome the democracy of it, the chance to see a change of scenery, to put some distance between yourself and the 'no problem' problem? Or does your inner ape tell you to bag a corner, place chairs around the nest and patrol the borders? The pros and cons of open-plan offices are hotly debated, and they come down to two issues: should you go ape or ant? The ant has the ability to move about according to taste or whim, forming loose groupings with people he needs at that particular moment. The ape's territorial tendencies usually create an egosystem where some executives bag the window seats, while others find themselves over by the photocopier, more or less a reflection of the pecking order.

A version of hot-desking is hot-train-seating, where the same commuters travelling up to town at the same time every day get into the habit of sitting in the same pattern. In one case reported to me the seating arrangements had become so habitual that when one of them was absent through illness, their seat remained empty, while other commuters were forced to stand.

Commuting

Commuting is fun, isn't it? Not! Crammed together in a train carriage, with the women on the same level as the men's armpits and the men up in the women's hair. The ape screams for release, feels threatened by the closeness of the others around, is too scared to speak, but stands quietly, avoiding eye contact. Unless the nation has won a World Series, World Cup or World War, when the whole ant-like flock is happy to share, and the carriage rocks and rolls all the way up to the Smoke.

Tea troll

Did you mistakenly make everyone a cup of tea a couple of years ago, and have you been making the tea ever since? It starts innocently enough – an act of generosity, and thanks all round. Do it again and the requests begin: could you make it weaker this time; one sugar please; I only have Earl Grey before eleven o'clock; my cup's the one with 'I'm a mug' on it. Quite soon it becomes clear that it would take too long to train anyone else. It's your job for life.

Has it generalized? Do people come to you when they want anything trivial done, tea-related or otherwise? Does your ant feel useful; does your ape feel used? It is good to be needed. It's good to keep busy, but at the end of a couple of years, what have you learned? How to separate teabags? This may not be in your future interest. Would you go so far as to get it wrong – make Lapsang Souchong for the Earl Grey lover or decaff for the coffee connoisseur? Will your inner ape fool them into thinking you are a lesser ant, just to shake off the chore?

Why not combine the two? With your best team-spirited smile, pin a rota to the kitchen cupboard and invite everyone to sign up.

The patsy

Are you the patsy? Are you the victim? Are you always to blame? You only have to be wrong a few times in your first week to earn the title. The role of patsy is an important one. As soon as it is allocated, the office can relax. Whatever goes wrong from now on can be laid at Patsy's cubicle door. More and more can be piled up there, and when there are staff cuts, Patsy will be the one who goes, leaving the others with a clean slate. There's very little Patsy can do. Being perfect from the second week onward is no defence. The first week cements the label into place. If anything goes wrong – whether or not Patsy was in the office at the time – all anyone has to do is shake their head in sorrow, glance in the direction of Patsy's desk and their fate is once again sealed. This is pure ape. Ants are incapable of such obsessive ranking and group bullying. In less competitive societies there is no scapegoat, no blame, no sackings, everyone has a value. Not so in business. How do you avoid being the patsy? On your first day when you join a team, look around to see who is the patsy. If you can't see the patsy, leave the team; you're the patsy.

Sartorial

Your new boss has a different line in clothing. The old boss was a suit-and-tie man, the new one favours linen jackets and open collars. Suddenly everyone else in the office has started wearing linen jackets and open collars. Is this 'ape *vs* ant' or are we talking sheep here? Well, it's a mix of ape and ant. The ant says all are equal, and one of the signs of equality is in the clothing. The ape looks to the leader for sartorial direction, which is why the fashion has changed with the boss.

Love

Office romance, it happens. It is frowned upon, but there you go, nothing can be done. Well, one thing can be done. You see, when love is in the air, it really is in the air. The ant's way of sending signals by pheromones, which we still use, is in top gear now. All the others in the room are getting a noseful of the passion. How confusing is that when you're going through a celibate phase? For goodness' sake, you two, shave your armpits and put on some antiperspirant!

Is it worse when the loved one is not in the office? Then, when Romeo phones Juliet, you get one half of the phone conversations, and have to work out the other half yourself.

Significance

Do you crave meaning in your job? Is it no longer rewarding? Are you frustrated? How strange. This is a very new phenomenon. In other lands and in other times, it was enough reward merely to get through life with lungs and limbs intact. People just existed together, doing whatever needed to be done, mucking in and playing the ant. But because we are trained in the ways of the ape there is a restlessness. Or is it the absence of disaster which unnerves us? We are, after all, evolved for catastrophe (see page 175).

The foundation of this is recognition. If you are not receiving as many gifts as the others you should consider giving more. In a perverse way this goes for work assignments as well as pink pandas. Leaving you with nothing to do is akin to edging you out (see page 170).

Nemesis

Your team has just taken delivery of a new young executive, keen, thrusting and effective. What do you do? Cry a lot? Is this upstart the end of your promotion prospects? Which part of you is telling you to throw things about the office, beat your chest, then fight to the death? Which part has the urge to just hunker down with the newbie, ant-like, warm and cuddly, showing him how the photocopier works, introducing the other team members? Make him feel at home, relax, share his ideas and dreams, express himself fully, so you can discover his weak spots and pummel them.

Won't do you any good, though; you're on the way out, Grandad. And so soon after your fortieth birthday, too.

Pay round

Is it time to ask for a rise? Do you think your work deserves a special extra bonus, considering the hours you put in and your natural talent? Or do you think everyone should be paid the same regardless of their contribution, because everyone is valuable in their own way? What are you, ape or ant? In nature, animals often share their food with members of the group who are injured, old or just unlucky. Elephants, bats, dogs, whales, among many others, perform basic Marxism – from each according to his ability, to each according to his need. Humans 'in the wild' are similarly philanthropic, ensuring everyone gets a share of the food, yet it seems impossible to practise such even-handedness in the corporate framework. This is unfortunate, because in every office at some stage there is a leak; the company's salaries are projected onto everyone's screens, to the shock, scandal and fury of everyone. Our innate sense of fairness goes into freefall. There are scratch-marks all the way up the walls.

Closing time

How are you with customers? The ones who turn up as you are closing the shop, plead they only want one thing, stop you from cashing up for three-quarters of an hour and then buy nothing. Can you stay smiling throughout? If so, then you are true to the ant within; if not, I pity the customer.

Decorum

Do people tell you what you would like to hear rather than what you *need* to hear? The politeness meme has its drawbacks. By being decorous people remain on speaking terms, but what they speak is not worth much. For those who truly wish to understand the tide of opinion within the department your best recourse is to eschew the verbal aspects of the communication. You can keep the iPod plugged into your ears, just look for the hidden meaning behind what they are saying to you – the body language, openness of gesture, the amount of smiling and the general tone of their voice. Forget the words. It might even help you to focus on the near-invisible micro-gestures and signals which tell you you're being sacked for listening to the iPod all day.

Catch-up

Is your company so far behind the times they still use foolscap paper? These days 'behind the times' can mean just 'last month' (as in 'You bought the computer *last month*? Well, I doubt you'll be able to get the parts any more. Pity, if you'd left it a week before buying, you could have got the latest version, neater and cheaper'). The rat race is fast and only the fittest survive, but while some companies are galloping, others are Galapagos. They were started up by gentlemen in 1979 who made a tidy sum of money and assumed the technology they were operating in those balmy days was the technology that would last a thousand years. (It is remarkable what an Amstrad 1640 can do, considering when it was launched in 1984, a megabyte of memory was a reckless dream). This is fine for the gents; soon they will retire and the business will blow away on the breeze. The problem is for you; should the ant in you say, 'Well, someone's got to be last', or does your ape yearn for better tools? Your inner ape has the choice; stay in the twentieth century or launch yourself into the present.

The C word

Are you constantly troubled by clients? Life would be so much easier if they didn't keep ringing all the time. For goodness' sake, they've given you the commission, now why don't they leave you alone to search for more trade? 'Client' refers to the company as a whole – Microsoft might be a client – but the person who is on the end of the blower is not the CEO, but a very, very junior executive who knows exactly what a dump-job they have, but thinks they can earn a few brownie points by demonstrating powers of command and control.

They command you because they own you, so you must be nice to them just as they must be nasty to you. They are apes and you must be the ant. Many companies, wise to this, insert a special weak link into the project in a place where it can be easily spotted by the feisty exec and easily corrected by you, to the satisfaction of all. One graphic artist of note lived in fear of the executive saying something ghastly like 'Could the sky be a little less cloudy' or 'I want the child with a balloon moved to just there'. So he got into the habit of drawing one of the characters just slightly wrong. The young executive scoured the picture for faults before pointing in triumph, 'That man has got six fingers!' The artist was ever so grateful; the executive was bathed in a glow of Ruskinesque glory; the finger was removed with one deft dab of a paintbush.

Record keeping
Are you in information overload? Everything is filed in several places – hard copy and copies of copies, including emails, memos and backs of envelopes. This is very good for instant access to information and insurance against server crash and nuclear war, but only for three months. After that there are so many drawers, box files and stacks piling up that the office needs underpinning with concrete piles to prevent the floor collapsing. Look on the good side; it's amazingly good exercise, heaving all those crates up from the basement and dragging them down again when they are found to be the wrong date.

Being right
Is the boss right even when he's wrong? Only the strongest, most confident manager can admit to being fallible. They are

the pivot and point of their team. Your reputation within the company depends on their image. You will protect their name, understate their blunders and celebrate their few moments of lucid thinking. It is important to realize the pressure there is on the whole team to bolster Boss, even when he's a complete twerp. Which is all very well, but if Boss comes to believe the bolster he is heading for captainitis.

For a manager who has stepped too far down the path of omniscience, the greatest confusion comes when he hires someone new, who then tells him he's stupid. This is one in the eye for omniscience. What a stupid fool was Boss to hire him. But, if he was a stupid fool, then the new boy was right, so it was a good decision to hire him, so Boss did the right thing, so that makes Boss wise again. But Newbie keeps telling him he's stupid, so Newbie must be stupid, but that means Boss shouldn't have hired him . . . This might take an afternoon to work out.

The second greatest confusion comes when Boss is eventually levered out of his position by the nemesis he himself hired. Away from the protection of his team, he finds everyone queuing up to tell him he's stupid.

Niceness

Is Ape-Manager suddenly being nice to you? It is strange to find yourself getting suspicious when people are friendly. However, if there is a change in behaviour, if a normally brutal boss starts hanging around you, alarm bells should ring. When the lion tries to lie down with the lamb, it may be because he wants the lamb to go spy for him.

Even the most brutal Ape-Manager comes to realize that a

team works because it wants to. However much you shout or bully, results don't happen unless you have them on your side, and try as you might, you can't beat friendliness into the team. Ape-Manager needs an ally, someone who can persuade the team to fall into line. If Ape-Manager is taking you aside for friendly chats, talking about your promotion prospects and the like, then you have been chosen to do his dirty work. It's time to work out – are you the ant who stays true to the team, or are you tempted by ape-like ambitions?

Appraisals

During appraisals should we be the ant and praise everyone, or should we allow our ape to lay everyone to waste? We should praise everyone, says the ant, even Vernice. Vernice was team leader, but spent her time carping at everyone's work. Our ant shows us a paradox; if you accuse Vernice of not being a team player you are yourself breaking rank and showing you yourself are not a team player. Fortunately, we have devised a secret code to show what we really mean; you damn her with faint praise. When they ask, 'Tell us how you got on with Vernice,' you leave a five-second pause before saying, 'I think Vernice is very talented. A very thorough and energetic person, who willingly shares her opinions among her team-mates and is generous with her advice.'

Exit interview

At last a chance to unburden oneself about the manager's hopelessly muddled messages, confidence-undermining sarcasm and exhausting bad jokes that have sapped your will to live for the last five years. Do you tell him now, here, for his own good, for the good of your colleagues and the company and for the good of the planet? Or do you go ant-like and say it was all fine? The temptation is strong to wave the ape's branch in a final fit of fury, but the ant in you says: be nice to people on your way up. You might meet them again on your way down.

Nothing to do

Do you have nothing to do? This may be cause to celebrate, but only if you are the Emperor of the Universe. Otherwise you should worry; they may have rumbled you. It is vital that you awake from your stupor and start to do your job, or something like it. Make work for yourself, otherwise at the next shake down your record card will be a blank. Perhaps there are others in the office who are keeping work away from you. Who knows, they may be doing it as a favour to you, or they may be making themselves indispensible, to be preferred for promotion, or they may have decided there's nothing they can trust you with. Check it out! The correct speed to be running at is the same as everyone else's; no faster and no slower. This is the way of the ant.

Are you expected to work all hours?

Most managers spend far too long at work, ruining their home life as a result. If you had any sense of solidarity you would too, so they think. The ant and the ape take turn and turn-about. If there's a crisis on – changed specifications, changed deadline, multiple bookings – then the ant says, 'Go on, ruin your own home life, just for a bit.' If there's no crisis, then there should be no panicking.

Caught in the middle

Are you middle management, being attacked by those above and those below? The people on the shop floor know what they are doing, and resent being overworked and underpaid by those who know nothing (that's you). The people on the top floor know what the accounts should look like, and resent a manager unwilling to go that little bit further to cut corners and reduce wages (you again). You should invite top management down to see what it's like in the engine room every now and then. Likewise, you should invite the toilers up to the boardroom, to help them understand what the company faces out there in the market.

Omniscience

Did you look around you one day and realize you are the only one there who really knows how to run the place? This is a moment to send a chill down your spine. If you see someone doing something the wrong way – I mean *really* wrong; the way you tried before and which cost the company dear, *that* wrong – you can either butt in and tell them how to do it properly, and while you're about it the correct way to make decisions, how to keep chairs from squeaking in that irritating way and the secret for a really good cup of tea, after all of which your manager will send you straight off for an early bath. Or you can bite your lip, till it bleeds if you have to, to stop you making things better. Businesses, like everything else, run on chaos. That is the ant's way. If there is only one way to do things, there is no adventure left for them.

The only recourse, if not to succumb to vast modesty or

alcohol, is to leave and become a consultant. Then you can be wheeled in during a crisis and wheeled out rapidly after.

Blushing

Do you blush all the time? Is it a source of embarrassment? Do you go pink at the mere thought of it? Why didn't blushing evolve out of the genome long ago? It's such a giveaway that you have done something wrong. But blushing is important. It is a social signal which indicates that you have somehow deviated from the standards expected by your group. We know this because of the exception that proves the rule: if you are singled out for special praise by the boss you blush, indicating a deviation above the group's norms.

Therefore, if you blush it shows that you have a sense of decorum. Some people blush a lot, some never blush. Who would you feel safest with?

SECTION 5

CONCLUSION

ALL CHANGE

Day by day the universe casually rips itself apart. Somewhere out there a pair of stars will collide tonight. When they do they will vaporize everything within a billion miles. (Not us because we're a trillion trillion miles away.) Just another day in the life of the cosmos. When catastrophes are happening on such a scale it's hardly surprising our wee planet is a bit of a shambles.

It tilts to one side, so the weather is different at one end of the year to the other. The molten rock at the centre keeps spilling out as volcanoes. Bits of crust (we call them continents) float around on the surface, bumping into each other and creating wrinkles; we call them mountain ranges. With added earthquakes, floods, droughts, hurricanes, blizzards etc. it's as well that animals keep evolving. They can't stay still when everything is changing about them. So the process by which two parents' DNA combines to make a baby is splendidly inefficient – every offspring is a mutant, slightly different from Mum and Dad. That's the way new species arise: by mistake, lots of mistakes. We are surrounded at the moment by evolutionary mistakes that worked.

We have adapted the way we produce things as technology and raw materials have changed. From the blacksmith's forge to nanotechnology, progress has been made through a string of errors. We are surrounded by technology mistakes that worked.

The big adaptation we have made is to a world that changes. Indeed, I believe we don't function very well without disasters. I believe we depend on them – we are catastrophiliacs.

CHAPTER 12

CATASTROPHILIA

It was Take Your Child To Work Day, and several of the office workers' little darlings were there. They were aged between seven and eleven and all, without exception, were bored. This was not a great introduction to the rest of their lives; all their mums and dads staring at screens full of text and telling them it was really exciting but could they stop fidgeting, please. But something was about to happen which made them think work was really cool.

Halfway through the morning, the ceiling above the filing cabinet crashed down and water came pouring in. Catastrophe! There was no time to dither, no time to think; everyone sprang into action. Immediately, the children were shepherded out to safety. Three of the staff hefted the filing cabinet out from under the deluge while another rescued the potted plant and pink fluffy dinosaur from on top. The office manager rushed out to alert the head of department of the disaster. Someone ran off to find a mop and bucket. Someone else went upstairs to deal with whatever was causing the flood. Others got down to turning off electrical appliances, snapping out the plugs and hauling desks to one side to clear a space around the leak. Within minutes the tap in the loo upstairs had been turned off and the wedge of toilet paper that someone's child had stuffed into the plughole had been removed, as had the child. The flooding stopped and everyone began to clear up the mess. Someone found a spare room nearby and everyone moved the files, computers and stationery there, plugged in, booted up, and when the office manager returned he found the team back at work, with the damp documents from the filing cabinet strung out to dry around their heads.

Later on the manager wrote a report to head office telling what happened, the state of the damage and what steps were taken. Head office sent him an email congratulating him on dealing with the crisis.

Two unremarkable details: one is that everybody mucked in. No orders were received from above, no requisition forms were filled in or protocols observed. Everyone did what had to be done. They each coordinated smoothly with each other, found a useful role to play and made a decent fist of it. The other unremarkable detail is that head office assumed the whole rescue was organized by the manager. Why should they not think that? It's what managers are supposed to do. But clearly, while the crisis was on, another authority was operating. With or without a manager, we can deal with disasters, because we are all catastrophiliacs.

Why is life so much more fulfilling when there's a catastrophe on? Is this the love that dare not speak its name? The slight twitch of delight when disaster strikes, the quickening of the pulse, the flaring of the nostrils? Why are we attracted to calamity rather than repelled? Why do we rush towards the danger zone rather than run away?

There is nothing odd here. It's what we are built to do. For aeons we have been surrounded by the regular crump of catastrophe. Earthquakes, tsunamis and volcanic eruptions come in thick and fast around the edges of the tectonic plates that form the Earth's crust. Under our feet the planet is skating around all over the place. The oceans are ripping apart down the middle. India is careering into Asia at the rate of one centimetre per year − not fast if you're a car, but impressive if you're a sub-continent. Even in the placid plateaus far from the geological firing line people have to cope with other catastrophes − drought, blizzards, famine, floods, fire and diseases − at least once a century. We have evolved on a capricious planet which will change its nature on a whim. If we were the least bit picky about our living conditions we would be extinct. As it happens, catastrophe is what we do best. After all, anyone can thrive during the good times; it's coping with disasters that mark us out as survivors or divers. We are all catastrophiliacs.

I offer thirteen proofs:

1) Proof by Japan

Japan is a particularly earthquake-prone country. Sitting on the 'ring of fire' that surrounds the Pacific Basin, there is a tremor of sorts just about every day (up to 1,500 every year nationwide), as the Pacific plate, the Philippine sea plate and the Amurian plate grind against each other. The plates have also ground their national character into a particular shape.

The Japanese have always fascinated the rest of the world. They are famously modest, respectful, deferential, unwilling to express their emotions openly and full of formalities that hedge their behaviour. From the outside they seem repressed, conformist, lacking individuality. They also have a very flat management structure: senior management are as respectful of their employees as their employees are of them. When a Japanese business goes bust we don't see the bosses stuffing their pockets with as much cash as they can, then disappearing to a far-away tax haven. They weep with shame publicly, sharing the grief of their employees. Because there may be an earthquake next week, they may be trapped under rubble, and they'll hope for rescue from one of the people they've just put out of work. Everyone helps everyone else because by helping each other they are helping themselves. This is true during a disaster, and in between disasters too.

Myrmecologists (ant experts) may think that the Japanese come close to the social structure of ants. They have one big thing in common: they are both used to catastrophes practically on a daily basis. Ants are forever dealing with floods, wayward gardeners' spades and the like – things that would be lethal to an individual ant. They have evolved to be 'eusocial' – entirely dependent on each other. There is no individual ant, just a colony. Ants would not survive if they didn't harmonize. Neither would Japanese people. Both ants and Japanese have adapted to be experts in catastrophe.

The Japanese may be a special case because catastrophe is a way of life, but everyone in a crisis can discover the instinct.

Instead of panicking or screaming, an unexpected sense of calm and sense descends. People forget about themselves and look for ways they can help those next to them. That simple act of looking after your neighbour turns a crowd of individuals into a web of communication and cooperation, for while you look after your neighbour, they are looking after their neighbour, who in turn is looking after theirs . . . and if you have been nice to everyone else, someone is looking after you.

2) Proof by war

During the Second World War, when London was about to be pummelled by enemy bombs in the Blitz, the Ministry of Defence drew up plans to cope with what they were sure would happen to the population – looting, murder, rape and public disorder. None of that happened. Instead, the public showed their famous British Stiff Upper Lip, kept calm and carried on. (A shop which had its windows blasted by a bomb displayed a sign saying 'More open than usual'.) But look around; that stiff upper lip of which the British are so proud is not a uniquely British ability; every nation on Earth does it when they are in similarly catastrophic conditions. So you haven't heard of the famous Turkish Stiff Upper Lip, or the famous Indian Stiff Upper Lip? They call it simply 'kismet' or 'karma'.

3) Proof by peace

Catastrophilia doesn't stop when peacetime starts. If by chance disasters don't seem to be happening right now, we feel uneasy. Politicians are cynically aware of this. Desperate to win votes at home, they will search for an enemy abroad, to keep the people busy doing what comes naturally – uniting against the threat, and therefore supporting the government. After the upheaval of the Russian Revolution, the Soviets imposed a string of Five-Year Plans, to turn the backward Soviet economy into an advanced industrialized state. They brought intense suffering, but hardship was tolerated because, as one put it, Soviet workers believed in the need for 'constant struggle, struggle, and struggle' to achieve a Communist society. Mao Zedong in China instigated the Great Leap Forward and the Cultural Revolution to keep the people in 'continuous revolution', a concept foreshadowed by George Orwell in his novel *1984*, in which Oceania was in a state of continuous war with Eurasia. When eventually a peace deal was struck with Eurasia, they promptly declared war on Eastasia on the other side.

4) Proof by bonding

Long after the Second World War people who had been through it would quietly mention that they really quite enjoyed themselves; got to meet their neighbours, bonded with their communities, forgot about family feuds and so on – sentiments helped of course by the warm glow of nostalgia and the useful additional fact that we won. Powerful, life-long bonds form between people who have been through hardships together. After five years of blood, sweat and tears in the Second World War, it seems perverse to say, as some did, that they were 'the best years of my life'. The old Chinese curse, 'May you live in interesting times', has multiple layers of irony.

5) Proof by science

The greatest advances in science happen during crises. Wartime of course produces many, as we have seen over the last century, when two world wars resulted in the accelerated development of nuclear power, radar, jet engines, nylon, artificial fertilizers, polythene, penicillin, anti-malarial drugs, aerosol sprays, caterpillar tracks, computers, tape recorders and rockets (and then, during the Cold War that followed, the GPS system, which was to search the ocean floors for enemy nuclear subs, but instead discovered continental drift and the cause of natural catastrophes: volcanoes, tsunamis, landslides and earthquakes).

Their rapid developments came from a massive sharing of ideas and endeavour borne out of the catastrophe of war. In peacetime we are used to a rather sedate progression of science and technology, where each new invention is accompanied by a careful construction of patents, legal fire walls and negotiations with manufacturers. Progress is decorous and slow – safety first. But during wartime there is no time to argue about copyright or primacy, form-filling, safety issues or permission-seeking. Ideas have to flow and change. Mistakes must be made in their hundreds and lives lost by the dozen to quickly get things right. Everybody's idea gets considered and everyone becomes a stakeholder in the result. Everybody is important, but nobody is more important than anyone else. For the evolution of technology the conclusion is inescapable: catastrophe is far from catastrophic – it's fertilizer.

6) Proof by art

The great art revolution of the European Renaissance
didn't happen during peaceful times, but throughout the
rivalries and skirmishes of the Italian city states, with
Rome, Venice, Florence and Milan deploying – in Raphael,
Titian, Michelangelo and Leonardo da Vinci – Weapons
of Mass Creation to proclaim their supremacy. Without
the revolution and danger that surrounded the rise of
Napoleon, it is doubtful if Beethoven's music would have
been so revolutionary and dangerous.

7) Proof by slum

There is a permanent flow of disaster around the great 'informal
settlements' – slums – of the world: Nairobi, Karachi, Mumbai
and many more across the globe, where nearly a billion people
live on the bread line. All the poverty, hunger, insanitary conditions and disease cannot grind them down. This is the natural
state of things, so they not only survive, helping each other and
sharing what little food they have, but visitors comment that
they are remarkably composed, accepting it as their lot.

8) Proof by proxy

Those of us who are far removed from any disaster much bigger than finding ants in the pantry may feel a hankering after
the bad times. News programmes and disaster movies feed
our catastrophiliac urges. In darkened rooms, with gloomy
glee we watch TV programmes on life-threatening diseases

and home make–overs. For the hardened addict there's disaster tourism, where you join busloads of day trippers, pop in on shattered towns and ruined lives, and take pictures for the folks back home.

9) Proof by manager

Managers have long understood that an air of imminent disaster is good for the workforce. The day after posting bullishly optimistic forecasts to the shareholders, they tell the employees they'll have to cut jobs and shorten lunch breaks. Call centres deliberately set one centre in rivalry against another. Like oarsmen on a galleon the operatives in Sunderland are whipped up to try and crank their sales rate to beat the infidel call centre in Warrington. Sometimes the only way to achieve peace at home is to be at war everywhere else.

10) Proof by allergy

We are encouraged to not wash our hands now, to eat unwashed vegetables, to let our children crawl around the garden to their hearts' content, then lick their hands all over, because that's the way to avoid allergies. Allergies and asthma happen when the immune system is taken by surprise by a new organic entity it encounters and flings up a hasty defence. The more we feed it new entities, the less it is surprised, the more robust we become. If we protect and cherish our immune system it will suffer from lack of danger and absence of challenge. Keep it under stress and you'll live long and prosper. What doesn't kill you makes you stronger.

11) Proof by diet

Obesity happens because we eat, and carry on eating even when we are embarrassingly fat; and beyond, when we are a threat to floors and furniture; and well beyond the capacity of our hearts and joints to bear our weight. Why? Because we are always expecting a catastrophe, and are frankly surprised when we don't find one. We carry on eating because any minute now the food will disappear, so we think.

12) Proof by males

The existence of the male on the planet has long been a mystery. In the past, animals and plants did quite well enough simply by podding. Asexual reproduction was the rule until about a billion years ago. The reason why things had to change was that diseases were continuously evolving new ways to attack, and creatures needed to rapidly evolve defences. Asexual reproduction wasn't a fast enough driver of change, genetically, because it produced near-identical offspring. Something was needed to shuffle the genetic pack, and that something was males. So males were invented to cope with the imminence of plague and pestilence.

13) Proof by deadline

The pocket panic we all carry around with us is the deadline. Nobody, but nobody, gets jobs done in good time for a deadline, unless they are not earthlings at all, but disguised scouts for an alien invasion. All must be completed in the final moments. As the saying goes: 'If it wasn't for the last minute, nothing would ever get done'.

CHAPTER 13

THE BALANCE OF APE AND ANT

The balance between ape and ant – self-interest and generosity – started four billion years ago. The story of early life was the survival of DNA in a bitter world, struggling on its own, selfishly. But the second part of its story is the discovery that working with others improved its chances of survival. The selfish gene succeeded through selfless behaviour. Cells began to go round in organized communities – multi-cellular organisms.

You and I are just such a community, with about 100 trillion cells working harmoniously together. We are so complex, with our array of kidneys, lungs, livers and guts coordinating with each other so smoothly, it's hard to see any connection between us and our bacterial origins. Yet our internal bits and pieces – lungs, kidneys, liver etc. – were probably free-floating organisms three billion years ago, which grouped together to be more effective, so starting the march towards the creation three billion years later of the wonder that is you.

It has been a long and interesting journey. Some would say an incredible journey. Some would say an unbelievably impossible journey, so to give an idea of how it happened let's take a look at an animal which evolved halfway from bugs to us.

The Portuguese Man-of-War: until recently we thought of it as a gloriously sophisticated and dangerous creature of the kind only Hieronymus Bosch could have devised. A jelly-like balloon, about as big as your head, floats at the sea's surface. Blobby and innocuous on top, underneath hangs a curtain of streamers and tendrils that look like party decorations, but have a much deadlier intent. The short strands are digestive organs waiting for a meal; the longer tendrils are going to find one. They trail behind the jelly, sometimes as far as 30 metres. Any fish that touches them triggers 1,000 deadly darts, which fire into it, paralysing it, while the tendrils stick to it and drag it upwards, back to the jelly. There it is dissolved by the digestive organs in a leisurely fashion.

We now know the Portuguese Man-of-War is not an animal, but a colony of several species living together cooperatively, in one of four divisions – flotation, stinging, digesting and reproduction. We know they are distinct species because each cell type has a free and independent cousin who swims in the ocean beyond. The independent creatures' genes are selfish, because when you're on your own keeping alive is a selfish kind of thing. But their cousins in the man-of-war have discovered that cooperation is more useful than selfishness, to the extent that they have given up whole stretches of their genes: they can no longer live independently of the mothership.

And now you have joined the man-of-war which is your workplace or community. The company you work for wevolved along similar lines to the man-of-war: to capture trade, digest the money therefrom and keep afloat in the marketplace. You help it do that, and you have to give up some of your independence in the process. You may have lost some diversity in your skills too, specializing in accounting, designing or marketing, for instance.

Being absorbed by the company can cause some pain and confusion at first. There is a feeling of loss of free will and independence. It takes time to adjust to company culture, but when it happens it can be surprisingly all-embracing, from the way you speak, the clothes you wear, even the tilt of your head. Philip Robinson advises new recruits at any company to keep a diary for the first year or so. The eccentric characters and rituals will seem strange in the early days, but when they look at the diary twelve months later they will wonder what they were making a fuss about. It will all seem quite normal.

The best companies have ants and apes in balance. The apes hunt the markets and fight for the best deals, the ants create the best products. But of course everyone is both ape and ant; for instance in brainstorming sessions, where everyone's ape challenges, defends, adopts and adapts ideas, while everyone's ants ensure agreement in the end.

What goes around comes around

Ilona Schofield was at the end of another day teaching in a secondary school in Brighton. A boy asked to borrow some money for his bus home. She gave him the money and he disappeared with it. Next day at the end of school he came up to her again. She reached for her purse again, but instead of asking for money again he gave her some. 'Here, Miss. Here's your money.'

'No,' she said. 'I didn't lend it to you, I gave it to you. Keep it. Just do something for somebody else some time.'

Ilona was subscribing to a kind of ideal, generous society where people help other people just because they need help. They don't ignore the call, nor do they help only if they get something in return. They altruistically give to anyone who needs it. If everyone behaved altruistically, everyone would benefit in the end. The three types of helping can be summarized in three charts of interconnectedness.

Firstly the selfish individual, in it for themselves; secondly the mutual back-scratcher, in it for close friends and family; thirdly the altruist, in it because we're all in it together. This, with all its wide connections, looks more like the 'fabric' of society.

If only this would work. If only everybody did behave altruistically. It is the dream of the monotheistic religions (including Judaism, Christianity, Islam, Sikhism, Hinduism). They agree God made us perfect, as an omnipotent deity should do, but they are bothered by the inconvenient truth that we often behave horribly to each other; we cheat, lie, bully and steal. Religions cannot explain selfishness. On the other hand, materialistic evangelists of the Selfish Gene cannot explain altruism. The answer we have found is that both are necessary. The antagonistic harmony between the two creates chaos. Without chaos there would be no change; without change no evolution; without evolution no life.

The see-saw between the two instincts was recognized by the Taoist philosophers 2,500 years ago in China. Taoists consider the whole universe to be dominated by complementary forces – action and non-action, light and dark, masculine and feminine. In the absence of modern scientific knowledge the ancient philosophy might seem lacking in depth, but that grasp of the fractal nature of the universe was incredibly perceptive. The Taoist symbol for yin and yang shows the interconnectedness of the ant/ape forces in the white and black swirls. The two smaller circles hint that within the yin, or ant, nature there are elements of the yang, or ape, nature. Likewise the ape side cannot thrive without a little bit of ant.

So when you are feeling claustrophobic; when the rotten decisions of those above you and the ignorance of those below and the inertia of all around drive you mad, don't worry; you are an ape, and very valuable. You drive change, prick complacency, inspire creativity. Carry on being critical, but make friends with your enemies, and help them with their projects, because you need them to care.

And when you are already spending too long in the office, and then you add managing the company sports club and Christmas party to your activities, and everybody is grateful but nobody lifts a finger to help, don't be disheartened; you are an ant, and so important that you should nudge everyone else to be as big-hearted as you, and mention it gently in public when they are not contributing to the general well-being. Because fairness is more important than politeness.

Then make them all a nice cup of tea.

APPENDIX

Introduction

1   Aesop (620–564 BC) was a fabulist or story teller. He himself may be as much a work of fiction as his stories. There is no hard evidence of his existence beyond a few scattered details of his life in ancient sources, including Aristotle, Herodotus, and Plutarch. An ancient literary work called *The Aesop Romance* tells an episodic, probably highly fictional version of his life, including the traditional description of him as a strikingly ugly slave who by his cleverness acquires freedom and becomes an adviser to kings and city-states. But many stories were gathered across the centuries and attributed to him in a storytelling tradition that continues to this day, of animals with human characteristics.

2   http://www.cam.ac.uk/research/news/necessity-is-the -mother-of-invention-for-clever-birds

Section 1: Harmony

Chapter 2: The science of harmonizing

1 Rizzolatti was engaged in describing the firing of neurons in the motor cortex from the point of view of input signals. He and his colleagues watched the pattern of firing neurons when their macaque monkey picked up a peanut. The first thing they noted was that before the capuchin moved its hand to pick up the nut its neurons fired the grasping pattern, in preparation for the action. But the neurons fired in the same pattern when the experimenters picked up the peanut, indicating that they had observed the movement and translated it directly into their own motor cortex, mirroring the firing of neurons in the experimenter's motor cortex. Since then the search has been on to find mirroring in other parts of the brain, with the expectation that where there is empathy, there will be mirror neurons. [http://www.scholarpedia.org/article/Mirror_neurons]

Rizzolatti G, Fogassi L, Gallese V. 2001. 'Neurophysiological mechanisms underlying the understanding and imitation of action'. Nat. Rev. Neurosci. 2:661–70

Rizzolatti G., Craighero L. 2004. 'The Mirror-Neuron System'. Annual Rev. Neurosci. 27 169–192

Chapter 3: Egology

1 http://news.mongabay.com/2005/0819-hippo_tortoise.
html – AFP 2005

2 The 'Prisoner's Dilemma' game is an iconic game which analyses selfishness versus cooperation, devised by Merrill Flood and Melvin Dresher working at RAND in 1950. Albert W. Tucker formalized the game with prison sentence rewards and gave it the name 'prisoner's dilemma' (Poundstone, 1992), presenting it as follows (from Wikipedia):

Two members of a criminal gang are arrested and imprisoned. Each prisoner is in solitary confinement with no means of speaking to or exchanging messages with the other. The police admit they don't have enough evidence to convict the pair on the principal charge. They plan to sentence both to a year in prison on a lesser charge. Simultaneously, the police offer each prisoner a Faustian bargain. If he testifies against his partner, he will go free while the partner will get three years in prison on the main charge. Oh, yes, there is a catch – if both prisoners testify against each other, both will be sentenced to two years in jail.

In this classic version of the game, collaboration is dominated by betrayal; if the other prisoner chooses to stay silent, then betraying them gives a better reward (no sentence instead of one year), and if the other prisoner chooses to betray then betraying them also gives a better reward (two years instead of three). Because betrayal always rewards more than cooperation, all purely rational self-interested prisoners would betray the other, and so the only possible outcome for two purely rational prisoners is for them both to betray each other. The interesting part of this result is that pursuing individual reward logically leads the prisoners to both betray, but they would get a better reward if they both cooperated. In reality, humans display a systematic bias towards cooperative behaviour in this and similar games, much more so than predicted by simple models of 'rational' self-interested action.

There is also an extended 'iterative' version of the game, where the classic game is played over and over between the same prisoners, and consequently, both prisoners continuously have an opportunity to penalize the other for previous decisions. The pattern of tit-for-tat emerges as the best strategy – i.e. if they betray me I'll betray them next time. If both players adopt this, they enter a pact of cooperation which earns them the best returns.

In iterated games where players can choose who they play with, or where they can punish betrayers, cooperative play quickly becomes the norm, since those who betray are shunned. (Proceedings of the National Academy of Sciences, vol. 100, p. 3531)

3 Christopher Boehm, *Moral Origins*. Basic Books 2012

4 Frans de Waal *Our Inner Ape*. Riverhead Books 2005

Section 2: Antagonism

Chapter 5: Ape and ant

1 Tschinkel W.R. 2002. 'The Natural History of the Arboreal Ant, Crematogaster ashmeadi'. 15pp. Journal of Insect Science, 2:12, Available online: insectscience.org/2.12

Section 3: Anatomy of Chaos

Chapter 9: Run-Away

1 Failing banks: http://en.wikipedia.org/wiki/List_of_bank_ failures_in_the_United_States_(2008–present)

2 http://www.michaeleisen.org/blog/?p=358

BIBLIOGRAPHY

Adams, Scott, *Random Acts of Management* (Boxtree, 2000)

Blackmore, Susan, *The Meme Machine* (OUP, 2000)

Boehm, Christopher, *Moral Origins* (Bantam, 2012)

Dawkins, Richard, *The Selfish Gene* (OUP, 2006)

Dutton, Kevin, *The Wisdom of Psychopaths* (Arrow, 2013)

Gladwell, Malcolm, *The Tipping Point* (Abacus, 2002)

Holldobler, Bert and E. O. Wilson, *The Superorganism* (Norton, 2009)

James, Oliver, *Office Politics* (Ebury, 2013)

Johnson, Steven, *Emergence* (Penguin, 2002)

Keller, Laurent and Elizabeth Gordon, *The Lives of Ants* (OUP, 2009)

Masson, Jeffrey and Susan Norton, *When Elephants Weep* (McCarthy, 2010)

Miller, Peter, *Smart Swarm* (Collins, 2010)

Parkinson, Northcote, *Parkinson's Law* (John Murray, 1958)

Peter, Laurence and Raymond Hull, *The Peter Principle* (Pan Books, 1971)

Reeves, Richard and John Knell, *The 80 Minute MBA* (Headline, 2013)

Robbins, Harvey and Michael Finlay, *Why Teams Don't Work* (Business Essentials, 2000)

Solnit, Rebecca, *A Paradise Built in Hell* (Viking, 2010)

Suroweicki, James, *The Wisdom of Crowds* (Abacus, 2005)

de Waal, Frans, *Our Inner Ape* (Granta, 2006)

Wilkinson, Richard and Kate Pickett, *The Spirit Level* (Penguin, 2010)